James Tyson, William F. (William Fisher) Norris

A Treatise on Bright's Disease and Diabetes

With Especial Reference to Pathology and Therapeutics

James Tyson, William F. (William Fisher) Norris

A Treatise on Bright's Disease and Diabetes
With Especial Reference to Pathology and Therapeutics

ISBN/EAN: 9783337183622

Printed in Europe, USA, Canada, Australia, Japan

Cover: Foto ©berggeist007 / pixelio.de

More available books at **www.hansebooks.com**

A TREATISE

ON

Bright's Disease and Diabetes.

WITH ESPECIAL REFERENCE TO

PATHOLOGY AND THERAPEUTICS.

By JAMES TYSON, A.M., M.D.,

PROFESSOR OF GENERAL PATHOLOGY AND MORBID ANATOMY IN THE UNIVERSITY OF
PENNSYLVANIA; ONE OF THE PHYSICIANS TO THE PHILADELPHIA HOSPITAL;
ONE OF THE VICE-PRESIDENTS OF THE PATHOLOGICAL SOCIETY OF
PHILADELPHIA; MEMBER OF THE COLLEGE OF
PHYSICIANS OF PHILADELPHIA, ETC.

WITH ILLUSTRATIONS.

INCLUDING A SECTION

ON RETINITIS IN BRIGHT'S DISEASE.

By WILLIAM F. NORRIS, A.M., M.D.,

CLINICAL PROFESSOR OF OPHTHALMOLOGY IN THE UNIVERSITY OF PENNSYLVANIA.

PHILADELPHIA:
LINDSAY & BLAKISTON.
1881.

TO

MY FRIEND AND COLLEAGUE

WILLIAM PEPPER, A.M., M.D.,

PROVOST AND PROFESSOR OF CLINICAL MEDICINE

IN THE

UNIVERSITY OF PENNSYLVANIA.

PREFACE.

It is quite usual for authors, in their prefaces, to give some reason for having written the book which they thus introduce, —in a word, to apologize for its appearance. In the present instance the existence of excellent treatises on both subjects covered by the title, make it more than usually difficult to justify the appearance of another. The writer can only say that, for more than fifteen years, his thoughts, his studies, and his practical work have all been in the direction suggested by these subjects, and that during that time material has passed under his observation which ought at least, if properly used, to have resulted in an experience which should be useful to others. Whether this has been the case or not, must be left to the decision of those who may first read or examine the work.

In carrying out his object, the author has necessarily, in order to insure completeness, availed himself of the work of others, as well as his own, endeavoring, however, in all instances, to give credit where credit is due, and if he has omitted any proper acknowledgment it has been entirely unintentional. In the section on the " Histology of the Kidney," while he has used the illustrations of other writers, he has also added a number of original drawings, from the careful pencil of Dr. George C. Piersol. It is thought they will add to the accurate knowledge of the subject.

The very wide difference in the nature and pathology of the two sets of diseases, has not been overlooked in including them under the same cover; but when it is remembered that although diabetes is not a urinary disease, it is nevertheless a disease to a knowledge of which a study of the urine is indispensable, and that one who devotes himself to the latter must inevitably be brought into frequent and intimate contact with diabetes, it is but natural that the consideration of the two conditions should be united.

1506, SPRUCE STREET, April 1st, 1881.

TABLE OF CONTENTS.

BRIGHT'S DISEASE.

SECTION   PAGE

I. STRUCTURE OF THE KIDNEY, ... 17

II. TESTING FOR ALBUMEN—ALBUMINURIA, ITS SOURCES AND MECHANISM OF ITS PRODUCTION, ... 50

III. CASTS OF THE URINIFEROUS TUBULES—THEIR NATURE AND CLINICAL SIGNIFICANCE, ... 62

IV. CLASSIFICATION OF BRIGHT'S DISEASE, ... 79

V. ACUTE PARENCHYMATOUS NEPHRITIS, ... 85

VI. CHRONIC PARENCHYMATOUS NEPHRITIS, ... 124

VII. LARDACEOUS DISEASE OF THE KIDNEY, ... 148

VIII. INTERSTITIAL NEPHRITIS, ... 165

IX. RETINITIS IN BRIGHT'S DISEASE, ... 192

X. SUPPURATIVE INTERSTITIAL NEPHRITIS, ... 200

XI. CYANOTIC INDURATION, ... 219

DIABETES.

I. DIABETES MELLITUS, ... 227

II. DIABETES INSIPIDUS, ... 298

LIST OF ILLUSTRATIONS.

COLORED PLATES.

Eye-ground in a case of diabetes, ... FRONTISPIECE
Waxy casts, some of them treated with iodine, . . OPPOSITE PAGE 68

WOODCUTS.

FIG.   PAGE

1. Longitudinal section through the kidney, pelvis of the kidney, and a number of renal calyces, ... 18
2. Partially diagrammatic sketch of the structure of the kidney in longitudinal section, ... 19
3. Polygonal appearance of the lobuli when cut transversely, . . 19
4. Diagrammatic representation of the course of two urinary tubules, 21
5. Diagrammatic exposition of the method in which the uriniferous tubules unite to form the primitive cones, ... 23
6. Section through the cortex of a fresh kidney, showing cloudy epithelium, ... 25
7. Transverse section through a convoluted tubule of a fresh dog's kidney, without the addition of any fluid, ... 26
8. Section of a convoluted tubule of a dog's kidney, ammonium chromate preparation, ... 26
9. Surface view of a convoluted tubule, ... 26
10. Isolated cells and rodlets from the rat's kidney, ... 27
11. An isolated fragment of a descending limb, showing the clear, delicate epithelium, with alternate projections caused by the nucleus, ... 27
12. Ascending limb of loop, showing imbricated arrangement of columnar cells, according to Ludwig, ... 27
13. a, a portion of a convoluted tubule; b, of an ascending limb of Henle's loop, according to Heidenhain, ... 27
14. Irregular tubule from the cortex of the kidney of the dog, . . 28
15. Section through cortical substance of a human fœtal kidney, showing epithelium covering glomerule and lining capsule, . . 31
16. Distribution of the larger bloodvessels of the kidney, . . . 32

FIG.   PAGE
17. Diagrammatic representation of the bloodvessels in the cortex of
      the kidney, ...   33
18. Diagram of the circulation in the kidney, ...   34
19. A part of a convoluted tubule from the newt's kidney, showing
      capillary vessels and nerve-fibres, ...   42
20. Longitudinal section through the cortex of the kidney,   .   .   43
21. Tangential section through the cortex of the kidney,   .   .   44
22. Transverse section through a papillary portion of a medullary
      cone, ...   45
23. Testing for albumen by nitric acid, ...   53
24. Blood-casts, ...   63
25. Epithelial casts, ...   63
26. Granular casts, ...   65
27. Oil-casts and fatty epithelium, ...   66
28. Pale granular and hyaline casts, ...   66
29. Hyaline casts, ...   67
30. Proliferation and thickening of the capsular epithelium with com-
      pression of the glomerule, ...   95
31. Desquamative glomerulo-nephritis, ...   96
32. Capillary loops, with proliferation of the nuclei,   .   .   .   98
33. Suppurative nephritis—cystic kidneys—impacted calculi, .   .   211
34. Course of glycosuric influence, ...   236
35. Course of glycosuric influence, ...   236
36. Fat emboli in bloodvessels of lung, ...   265

BRIGHT'S DISEASE.

SECTION I.

STRUCTURE OF THE KIDNEY.

THE few facts in the coarser anatomy of the kidney which it is necessary to recall with a view to a correct understanding of its diseases, are, before section of the organ, its size, weight, color, consistency, and the relation of its capsule to its substance; after section, the appearance of the cortex or convoluted portion, as contrasted with the pyramids of straight tubes in the medulla, and the relative area of each.

As to *size* and *weight*, the adult kidney is usually about 11 centimeters (4.4 inches) in length, 5 centimeters (2 inches) wide, and .75 centimeter (.3 inch) in thickness. It weighs in the male 113.5 to 170 grams (4 to 6 ounces); female, a little less, 113.5 to 156 grams (4 to 5½ ounces). Its *color* in health is dark red, *surface* smooth, and in *consistence* it is usually firm and slightly elastic. The *capsule* is easily stripped off from the substance of the organ, dragging none of the proper glandular structure with it.

On *section* of the kidney the *cortex* is found to be granular in appearance, and uniformly light-red in color. It varies somewhat in width, but is usually 5 to 6 millimeters (⅕ to ¼ inch) wide, and in longitudinal section is seen to dip down between the pyramids of the medulla. These pyramids, ten to fifteen in number, are striated or fibrous in appearance, uniformly dark-red, and terminate in as many papillæ in the pelvis of the organ.

2

More important to a correct understanding of the pathology of kidney diseases is a knowledge of the minute structure of the organ. Even the naked eye can discover further differences on the surface of a longitudinal section of the kidney, such as is

FIG. 1.

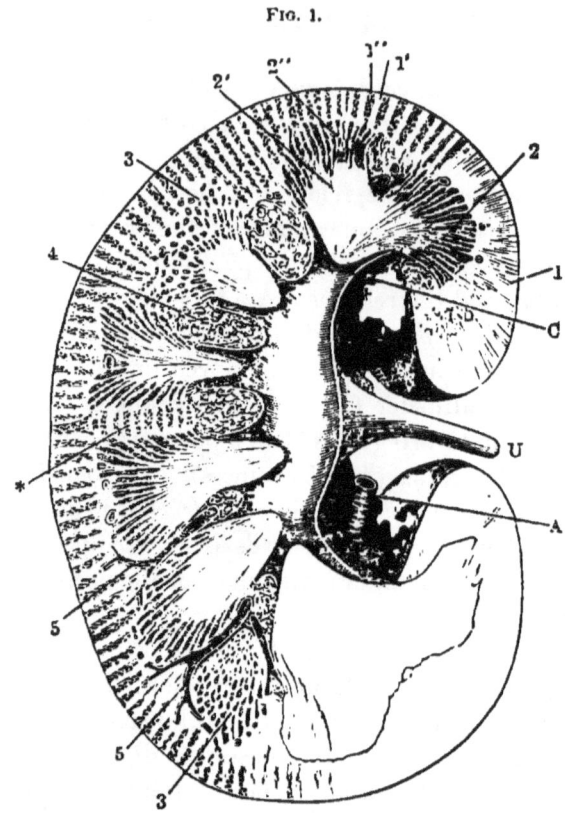

Longitudinal section through the kidney, the pelvis of the kidney, and a number of renal calyces. A, branch of the renal artery; U, ureter; C, renal calyx; 1, cortex; 1', medullary rays; 1'', labyrinth, or cortex proper; 2, medulla; 2', papillary portion of medulla, or medulla proper; 2'', border layer of the medulla; 3, 3, transverse section through the axes of the tubules of the border layer; 4, fat of the renal sinus; 5, 5, arterial branches; * transversely coursing medullary rays.—After HENLE.

presented in Fig. 1; and these differences are rendered still more striking if the bloodvessels are injected with a coloring matter. From each papilla as a centre radiate the excreting tubules, which appear as dark-red striæ in the natural organ. For a short distance they remain in contact, forming a continuous surface, known as the *papillary portion* of the medulla (Ludwig), or the *medulla proper* (Henle). This is represented at

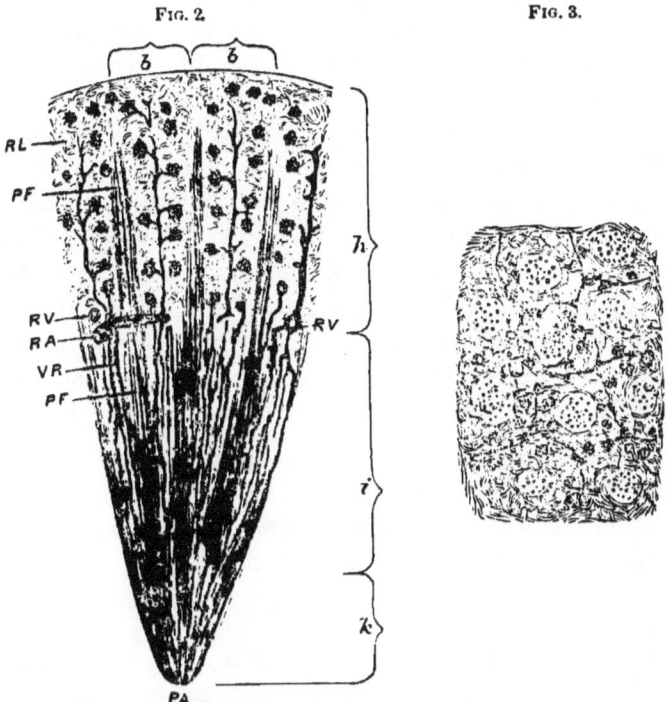

FIG. 2.   FIG. 3.

FIG. 2.—Partially diagrammatic sketch of the structure of the kidney in longitudinal section, and, Fig. 3, Tangential section through the cortex; bloodvessels injected. After Rindfleisch and Ludwig, modified. *k*, papillary layer; *i*, border layer; *h*, cortex. The lighter striæ, *PF*, are bundles of uriniferous tubules, a part of which are seen prolonged into the cortex as medullary rays or pyramids of Ferrein. Between each two of these in the cortex is the *renal labyrinth*, or secreting portion proper of the kidney. *a, a, a*, embrace the bases of the renal lobules, which, in transverse section at Fig. 3, appear as polygonal figures. *RA*, a larger branch of the renal artery, which separates the medulla from the cortex; *RV*, lumen of a renal vein which takes up the interfascicular vessels; the latter appear in transverse sections on the surface as stellate figures; *VR*, straight vessels (*vasa recta*); *PA*, surface of a renal papilla.

FIG. 3.—Polygonal appearance of the lobuli when cut transversely. *F*, transverse sections of the tubules forming the pyramids of Ferrein.

2′, Fig. 1, and at *k*, Fig. 2. As they proceed towards the periphery, the striæ become separated into bands of nearly equal width, between which are collections of other striæ. These in an injected kidney exhibit the color of the injecting fluid, but in an uninjected organ, though lighter or less opaque, are scarcely distinguishable from the bundles of straight tubes with which they alternate, unless they happen to be filled with blood. These striæ are, therefore, bloodvessels, and the portion of the medulla which is thus made up of alternate bands of bloodvessels and straight tubules is called the *boundary layer*, or *marginal layer* of the medulla (2″, Fig. 1, and *i*, Fig. 2). Close examination, especially with a low magnifying power, enables us to trace into the cortex almost to its border the bundles of striæ or straight tubes which come from the papillæ (1′, Fig. 1, and *PF*, Fig. 2). In this situation they are called *medullary rays*, or pyramids of Ferrein. Alternating with these medullary rays in the cortex is a *granular* portion, yellowish-red in the uninjected organ, which is the *renal labyrinth*, or *cortex proper*.

The further study of the minute structure is best facilitated by a separate consideration of the *tubules*, *bloodvessels*, and *connective tissue* elements, and the *lymphatic* and *nervous* elements which accompany them.

I. *The Uriniferous Tubules.*

These, with the bloodvessels, make up the great bulk of the organ. They may be studied from their beginnings in the cortex, or backwards as it were, from their terminations in the discharging-tubes on the papillæ. Selecting the former, the little granules visible to the naked eye in the labyrinth are the beginnings of the uriniferous tubules. These are spherical dilatations, .15 to .2 mm. ($\frac{1}{160}$ to $\frac{1}{125}$ inch) in diameter, formed of basement membrane lined by a mosaic of pavement epithelium. They are the so-called Malpighian *capsules* (1), Fig. 4, and are continuous, by a necklike construction (2), with the *proximal* convoluted tubule (3), which winds towards the adjacent medullary ray; reaching which, it passes vertically downward as the *spiral tube* (4) of Schachowa. These portions are all in the cortex, A. At the junction of the cortex and the border layer, the spiral tubule becomes suddenly narrower, and dips down through this layer, B, as the *descending limb* (5) of Henle's loop, the loop itself (6) being formed in the

Fig. 4

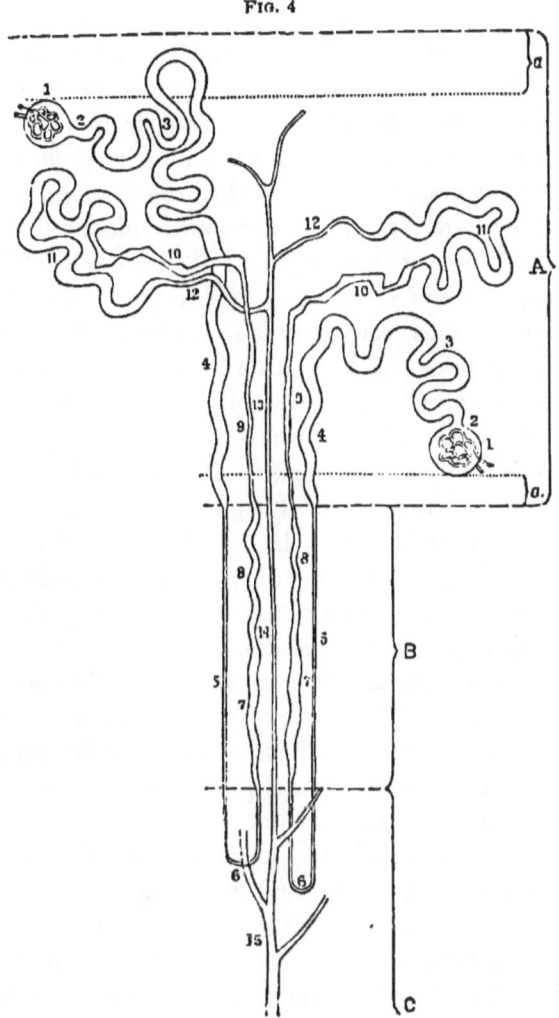

Diagrammatic representation of the course of two urinary tubules.—After KLEIN and NOBLE SMITH.

papillary layer, C. Re-entering the boundary layer, the tube becomes suddenly enlarged and slightly wavy in its course, forming "the first thick portion" (7) of the ascending limb of

Henle's loop. At the middle of the boundary layer it again
becomes narrower and spiral in its course, forming the "spiral
part" (8) of the ascending limb. The ascending limb now
re-enters the cortex in the medullary ray, again becoming narrower, though not of uniform diameter, and straighter, but
still slightly wavy in its course (9). Sooner or later, after re-
entering the cortex, the tubule dips away from the medullary
ray and passes into the labyrinth as a tube irregular in outline,
breadth, and course, whence it is called the *irregular tubule* (10).
Fairly in the labyrinth, it again becomes wider and convoluted,
forming the intermediary segment (Schaltstück) of Schweigger-Seidel, or *distal* convoluted tubule (11). Finally, by an
archlike turn, whose convexity is towards the circumference of
the kidney, it passes back towards the medullary ray as a narrower "curved part" (12) of the collecting-tube, which, uniting
with other similar tubules, forms the "straight part" (13) of
the collecting-tube in the cortex. The collecting-tube passes
down, still as a straight tube (14), though growing gradually
wider, through the boundary layer. In the papillary layer it
becomes the large collecting-tube, or tube of Bellini, which,
uniting with others, forms the "ducts" or "discharging-
tubes" which open on the papilla.

In its course from its commencement in the Malpighian capsule to its termination in the collecting-tube the uriniferous
tubule changes its diameter many times. First, there is the
constriction at the neck of the capsule, which is less than one-
half the diameter of the convoluted portion immediately succeeding it, and of which the average width, in the adult, is
about .06 mm. ($\frac{1}{400}$ inch), and in the newborn child, about
.016 mm. ($\frac{1}{1500}$ inch). The third change is found in the narrowing as the tubule becomes the descending limb of the loop
of Henle. The fourth is the slight dilatation just before or
after the loop is formed. Fifth, a slight temporary narrowing
at the "spiral part" of the ascending limb. Sixth, a further
narrowing as the ascending limb re-enters the cortex. Seventh,
the numerous changes in the irregular tubule (10). Eighth,
the widening of the intermediary portion, of which the diame-

ter about equals that of the convoluted tubule. And, ninth, a constriction as the latter terminates in the *collecting-tube.*

After the collecting-tube is formed it receives, just below its summit, a few additional canals, and then passes singly down to the papillary part of the medulla. Its diameter in the medullary ray is between .04 mm. ($\frac{1}{625}$ inch) and .08 mm. ($\frac{1}{312}$ inch) in the adult, and in the child .008 ($\frac{1}{3125}$ inch) and .05 mm.

Fig. 5.

Diagrammatic exposition of the method in which the uriniferous tubes unite to form primitive cones.—After Ludwig.

($\frac{1}{500}$ inch), the narrowest canals being in the centre of the ray. Having reached the papillary portion, the collecting-tubes unite by twos; first those of a single medullary ray, forming the principal or excreting tubes; then these unite with other excreting-tubes to form the papillary ducts, of which fifteen to twenty open on the surface of each papilla into the pelvis of the kidney.

Dr. George H. Rose, in an able graduation thesis *On the Arrangement of the Uriniferous Tubules,* presented to the Medical Faculty of the University of Pennsylvania, in March, 1879, concluded, first, that not all of the tubuli uriniferi go to form loops of Henle; second, that the medullary rays are made up of the convoluted tubes which do not form loops, as well as the collecting-tubes and ascending and descending limbs of Henle.*

Structure of the Uriniferous Tubules.—The Malpighian capsule is composed only of the basement membrane and mosaic of epithelial squams to which allusion has already been made. The cells, as seen in nitrate of silver preparations, are relatively large, although not uniform in size, and are provided with round and oval nuclei. The nuclei are arranged in groups of from two to ten. The group nuclei are round, about .01 mm. ($\frac{1}{2500}$ inch) in diameter, and, though flat, project slightly into the cavity of the capsule (*d*, Fig. 15). Each is surrounded by an amount of protoplasm, which is smaller the greater the number of nuclei in a group, whence the varying size of the cells. In the young kidney these cells are less flattened, and the nuclei project still further into the cavity of the capsule, as is seen in Fig. 15.

The capsules thus formed surround the capillary ball or glomerulus presently to be described, forming with it the Malpighian *corpuscle*. The latter are encircled by a few concentric layers of connective tissue, which are most numerous about those nearest the medulla. (See Fig. 15.)

This single layer of cells lining the capsule, which is comparable to an endothelium rather than an epithelium, extends into the neck of the capsule, whence onward to the papillary portion of the medulla the uriniferous tubule is composed of a basement membrane lined with an epithelium. The basement membrane contains an occasional nucleus, and in nitrate of silver preparations the convoluted tubules may be imperfectly mapped off into endothelial plates like the walls of blood and lymph vessels similarly treated; but the membrane is otherwise quite glasslike, affording, when the epithelium is washed out, one of the most satisfactory and typical examples of a homogeneous membrane.

---

* Dr. Rose's studies were made with extreme care upon macerated preparations of the kidney. They were many times repeated, and his thesis received the award of a prize.

The *epithelium* below the neck varies greatly, although it is everywhere single-layered and nucleated. The nucleus, alone, of the various cells is quite uniform, being round, sharply defined, and slightly granular. On the other hand, in the convoluted part of the tubule, the protoplasm of the adjacent cells is not differentiated, but each cell *fuses*, as it were, into its neighbor, so that the lining of the tubes, especially when acted upon by acetic acid, presents the appearance of a *nucleated protoplasm* rather than of a number of separate nucleated cells. At intervals may be observed, however, more especially if the tubules have been injected, certain

FIG. 6.

Section through the cortex of a fresh kidney, showing cloudy epithelium. The spheroidal nuclei are concealed. In the wider tubules irregular, in the narrower, regular fissures divide the epithelial mass.——After LUDWIG.

clefts or fissures where the cell bodies are not thus completely fused. These are well shown in Fig. 6, from Ludwig. According to Klein, the clefts are occupied by minute septules, containing here and there an oblong or angular nucleus, which pass in from the basement membrane. Of importance in connection with the study of diseases of the tubules is the fact that, in this portion, the epithelium is but loosely attached to the basement membrane, and may even be pressed out in the shape of a solid epithelial cylinder, or left as such by retraction of the basement membrane. The protoplasm of the cells in this situation is more or less "cloudy" in health, from the presence of a number of dark granules of albuminous composition, and even of some minute fat-globules. These serve to render the nucleus more or less obscure, while the addition of acetic acid, dissolving the albuminous granules, renders it again distinct.

R. Heidenhain* adds to the structures as above described an

FIG. 7.　　　　　　FIG. 8.　　　　　　FIG. 9.

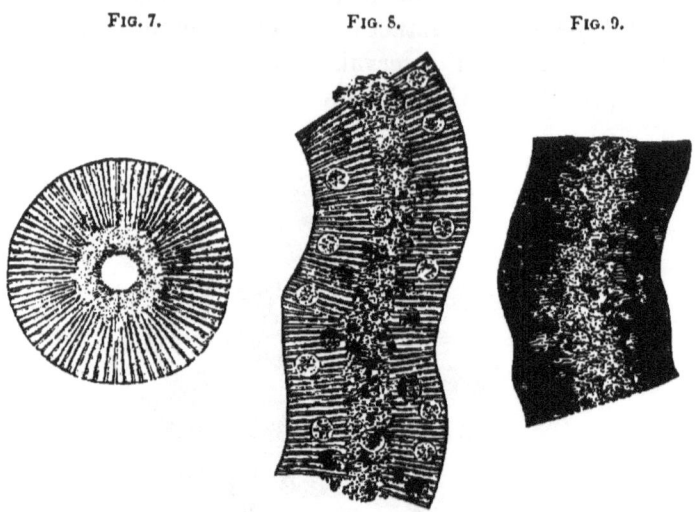

FIG. 7.—Transverse section through a convoluted tubule of a fresh dog's kidney, without the addition of any fluid. Hartnack imm. ix, tube in, drawing prism. Magnified 500 diameters.—After HEIDENHAIN.

FIG. 8.—Section of a convoluted tubule of a dog's kidney, ammonium chromate preparation. Hartnack Obj vii. Drawing prism, tube out. Magnified 440 diameters.—After HEIDENHAIN.

FIG. 9.—Surface view of a convoluted tubule. Same treatment and amplification.—After HEIDENHAIN.

additional element, the effect of which is, as he says, to make the epithelial cell of the convoluted tubules a very complicated organized structure. According to him the protoplasm is further differentiated into a large number of delicate cylindrical structures, which he calls "rodlets" (Stäbchen). These, rest-

* R. Heidenhain, Mikroskopische Beiträge zur Anatomie und Physiologie der Nieren, in Max Schultze's Archiv für Mikroskopische Anatomie. Bd. x, 1874, s. 1.

ing their peripheral ends on the basement membrane, perforate the epithelial layer in a radiated manner, imbedded in a very small quantity of formless matrix. The rodlets envelop, "like a mantel," the nuclei, which are placed at regular intervals, surrounded by a considerable remnant of undifferentiated protoplasm. According to Heidenhain the minute granules, for-

FIG. 10.   FIG. 11.   FIG. 12.   FIG. 13.

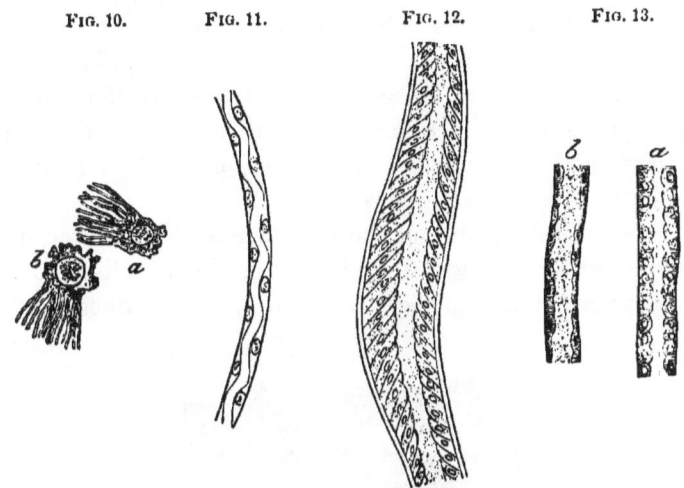

FIG. 10.—Isolated cells and rodlets from the rat's kidney. Same treatment and amplification.—After HEIDENHAIN.

FIG. 11.—An isolated fragment of a descending limb, showing the clear, delicate epithelium, with alternate projections caused by the nucleus.—After LUDWIG.

FIG. 12.—Ascending limb of loop, showing imbricated arrangement of columnar cells, according to Ludwig.

FIG. 13.—a, a portion of a convoluted tubule; b, of an ascending limb of Henle's loop, according to Heidenhain. The ascending limb has the greater lumen. Klein confirms Heidenhain in all essential particulars, but says, also, that, in the narrow part of the ascending limb, about the middle of the boundary layer, "many cells appear possessed of short processes and are more or less imbricated, a fact already known to Steudener."

merly described as present in the matrix of the cells, are nothing but the transverse sections of the rodlets.

Figs. 7, 8, and 9, from Heidenhain's paper, exhibit the appearances described.

Dr. Klein, of London, substantially confirms the observations of Heidenhain, but says these rods or fibrils, when looked at from the surface, are clearly connected into a network, so that they are more probably "septa of a honeycombed network seen in profile." In the *Atlas of Histology*, part xi, Fig. 1, recently issued by Dr. Klein, in connection with Mr. E. Noble Smith, this peculiarity in the structure of the epithelium is exquisitely brought out.

In the narrow descending limb of Henle's loop is found a single layer of epithelium, the cells of which are characterized by a transparent protoplasm and oval nucleus, which projects into the lumen of the tubule at alternate points, as shown in Fig. 11. The resemblance of these tubules to capillary blood-vessels is very marked, especially in transverse section; but, as Klein points out, the nuclei are more numerous than in a capillary, and the tubule has, further, a *membrana propria* outside the layer of cell-plates.

In the ascending limb of the loop, where the diameter again becomes greater, Ludwig describes the epithelial cells as columnar, arranged in an imbricated manner, and distinctly differ-

FIG. 14.

Irregular tubule from the cortex of the kidney of a dog.—After KLEIN.

entiated by a cleft about half way between two nuclei, as in Fig. 12. But according to Heidenhain's investigations, the epithelium of this broader ascending limb agrees essentially with that of the convoluted portion of the tube, the rodlets being similarly developed, as shown in Fig. 13. It is, however, tolerably easily distinguished from the convoluted tubule by the fact that the lumen of the latter is much narrower in consequence of the greater thickness of its cells, which, therefore, encroach more upon its lumen.

The irregular tubule, as figured by Klein, Fig. 14, is a striking object. He describes it as follows:

"It is situated among the convoluted tubes of all parts of the labyrinth; it has a very irregular and angular outline; in some places three and four times as thick as in others, and then of almost the same breadth as a convoluted tubule. But this greater thickness is due merely to a greater thickness of the lining epithelial cells, the lumen remaining everywhere a narrow canal. The cells show, in the profile view, exceedingly thick rods of a bright, homogeneous aspect, more distinct than in any other section of the urinary tubule. The cells are very angular and, in many places, imbricated; but in this, as in the former cases, this imbrication is due to the irregular outline of the tube and the great variation in height of the adjacent cells. Each cell possesses an oval or angular nucleus next the lumen."

In the intermediary portions the epithelium reassumes the appearance of that of the convoluted segment of the tubule; and here, again, Heidenhain and Klein reintroduce the rodlets, or, rather, their ends, as the cause of the faintly granular appearance.

In the collecting-tubes the epithelium is described as columnar, but the cells are more nearly cubical; they are sharply defined, rather broader at the base, which is seated on the wall of the tubule, while their truncated apices are directed towards its lumen. In the smaller canals, which unite in the medullary rays to form larger tubes, the round nuclei are surrounded with but a small amount of protoplasm. In the larger tubes the protoplasm is more abundant and the nuclei about the same size as in the smaller. Isolated cells from both regions exhibit projections and prolongations of protoplasm, so that it is really only in transverse sections of the collecting-tubes that the columnar or cubical appearance of the cells is maintained. In the papillary ducts a basement membrane can no longer be differentiated from the connective tissue around it, and the cells, which are here more elongated and typically columnar, rest upon the connective tissue alone.

Heidenhain says that, in the primary branches of the collecting-tubes, he has, here and there, observed that the so-called cylindrical cells are placed in an oblique manner, so as to give an imbricated appearance, and suggests that this may have given rise to the view of Ludwig, already described in these pages, that the ascending broad limb of Henle's loop is thus lined.

Anatomical Peculiarities of the Glomerulus or Malpighian Tuft; its Epithelial Covering, etc.—It is now generally conceded that, in addition to the mosaic of epithelium which lines the Malpighian capsule, the glomerulus itself is covered with a layer of epithelial cells, but all observers are not agreed as to its exact arrangement. According to some it simply covers the surface, bridging the space between the capillaries and the lobules which compose the glomerulus. Others consider that it also dips down between the capillary loops. The studies of Heidenhain[*] and of Langhans,[†] which are among the most recent, lead them to accept the latter view, while Schweigger-Seidel[‡] and Von Seng[§] adopt the former. The cells are for the most part well defined, but differ somewhat according as they cover the surface of the glomerulus or dip down between the capillary loops. In the former instance they are thin vaulted plates, whose concave surface rests upon the convexity of the capillary loop and is moulded to it. The shape of the cells also varies somewhat, according as they cover one or more capillary loops. In the latter instance the under surface of the cell is divided by projecting edges into several concave areas for the reception of the individual capillary loops. The edges unite at the centre, at which is the nucleus, which itself is conical in shape. At this situation—the seat of the nucleus—the external or upper surface is frequently depressed, so that the lateral extensions of the cell assume a pterygoid appearance. Other cells are concave on both surfaces, an effect probably due to their being insinuated between two capillary loops.

The nuclei measured by Langhaus are oval, .01–.014 mm. ($\frac{1}{2500}-\frac{1}{1780}$ in.) long and .006–.01 mm. ($\frac{1}{4188}-\frac{1}{2500}$ in.) wide, and somewhat flattened. The cells themselves reach .025 mm.

---

[*] Heidenhain, Archiv für Mikrosk. Anatomie, B. x, 1874, p. 3.

[†] Langhans, Ueber die Veränderungen der Glomeruli bei der Nephritis nebst einigen Bemerkungen über die Entstehung der Fibrincylinder. Virchow's Archiv, 76 Bd., erstes Heft, April, 1879, p. 85.

[‡] Schweigger-Seidel, Die Nieren, Halle, 1865, p. 76.

[§] Victor von Seng, Ein Beitrag zur Lehre von den Malpighischen Körperchen der Menschlichen Nieren. Wiener Sitzungsberichte, lxiv, 13 April, 1871.

($\frac{1}{1000}$ in.) in diameter at a maximum.   In some instances the line of junction between neighboring cells is not demonstrable, so that the covering of the glomerulus appears as a homogeneous nucleated plate.   The cells and nuclei are much more easily demonstrable in the fœtal kidney, in which they are more cubical, as seen in Fig, 15, from Klein's Atlas; undergoing a change after birth which is compared by Schweigger-Seidel to the post-fœtal transformation of the alveolar epithelium of the lung.   The epithelial covering of the glomerule is con-

FIG. 15.

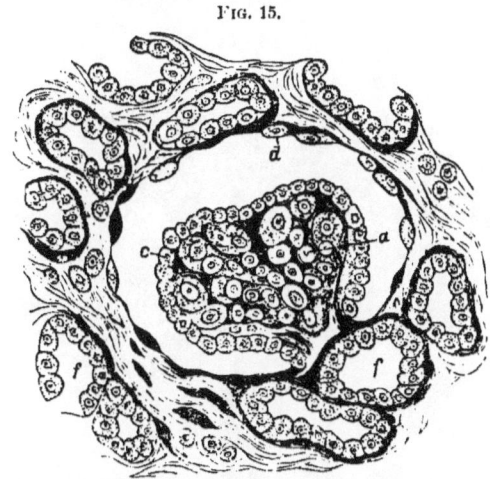

Section through cortical substance of a human fœtal kidney. *a*, glomerulus ; *c*, epithelium covering the glomerulus ; *d*, flattened epithelium lining Bowman's capsule ; *f*, urinary tubules in section.—After KLEIN and NOBLE SMITH.

tinuous with the epithelial lining of the capsule, as is well shown in Fig. 15.

According to Axel Key the capillary bloodvessels of the lobules of the glomerules are held together by a homogeneous connective tissue containing flattened stellate and nucleated connective tissue corpuscles.   But Langhans was unable to convince himself of the presence of connective tissue, except the adventitia of the vas afferens, which extends only until the vessel begins to break up into capillaries.

The capillaries of the glomerule or Malpighian tuft, like all capillaries, are provided with nuclei, which in common with all the cellular elements named are subject to pathological change.

There are no Malpighian corpuscles in that part of the cortex immediately bordering the capsule, nor in that bordering the boundary layers, that is, in the most external and most internal portions.   See *a* and *a*, Fig. 4.

II. *The Bloodvessels of the Kidney.*

The renal artery, before it enters the substance of the organ,

FIG. 16.

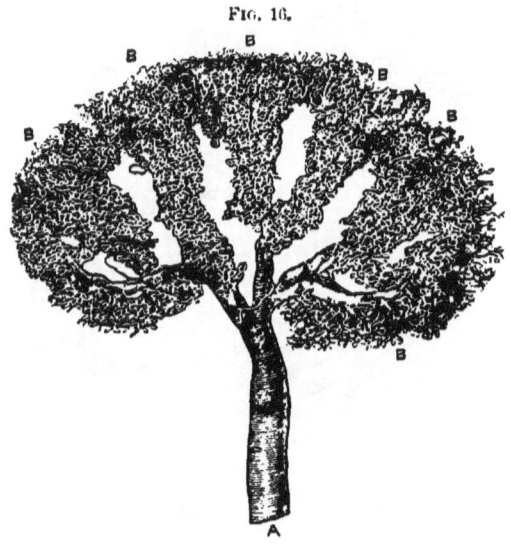

Distribution of the larger bloodvessels of the kidney. Engraved from a photograph of an injected and macerated preparation by Hyrtl in the Museum of the College of Physicians, Philadelphia. *A*, the artery. *B, B, B, B* correspond to the distinct lobar divisions of the kidney, to each one of which the branches of an artery are distributed, which again unite to form the area of the cortex. The open spaces correspond to the areas of the straight tubules or pyramids, which containing comparatively few blood-vessels were washed away in the maceration. The vertical interlobular arteries alluded to in the text are not seen, because the kidney is shown entire and not in hemi-section.

breaks up into two or three branches, which penetrate the capsule at the border of the pelvis of the kidney, and, after further rapid subdivision, radiate directly to the upper part of the marginal layer of the medulla.   From branches arching more or less irregularly along this border ascend vertical branches.

These pass into the labyrinth midway between two medullary rays, and are called interlobular or interfascicular arteries. From them pass off, laterally, numerous smaller branches called afferent,—*vasa afferentia*,—each one of which promptly perforates the nearest Malpighian capsule at a point opposite to the constricted neck of the latter. The vas afferens, according to the measurements of Thoma,* is .014 to .02 mm. ($\frac{1}{1785}$ to $\frac{1}{2575}$ inch) in diameter. Having perforated the capsule, it immediately splits up, forming a capillary ball—the *Malpighian tuft* or *glomerule*. The capillary vessels of this ball reunite within the

FIG. 17.

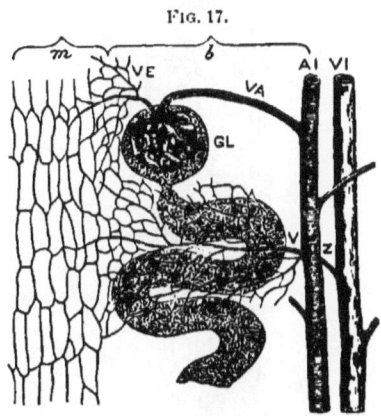

Diagrammatic representation of the bloodvessels in the cortex of the kidney. *m*, region of the medullary ray; *b*, region of the tortuous portion of the tubules; AI, interlobular artery; VI, interlobular vein; VA, vas afferens glomeruli; VE, vas efferens; GL, Malpighian capsule containing in its interior the capillary ball known as the glomerule, or Malpighian tuft; VZ, venous twig of the interlobular vein.—After LUDWIG.

capsule to form a vessel,—the *vas efferens*,—which perforates the capsule outwards at the same point at which the vas afferens enters, and slightly exceeds the latter in diameter. The efferent vessel proceeds towards the nearest medullary ray, and becomes a second time capillary. The resulting network embraces partly the straight tubules of the medullary ray, and partly the tortuous tubules, and, in the extreme periphery of the cortex, to which the medullary rays do not extend, the tortuous

* Thoma, Zur Kenntniss der Circulationstörung in den Nieren bei chronischer interstitieller Nephritis. Virchow's Archiv, Bd. 71, 1877, s. 65.

tubules only. To these capillaries are added others, derived from a few minute branches given off by the afferent vessels

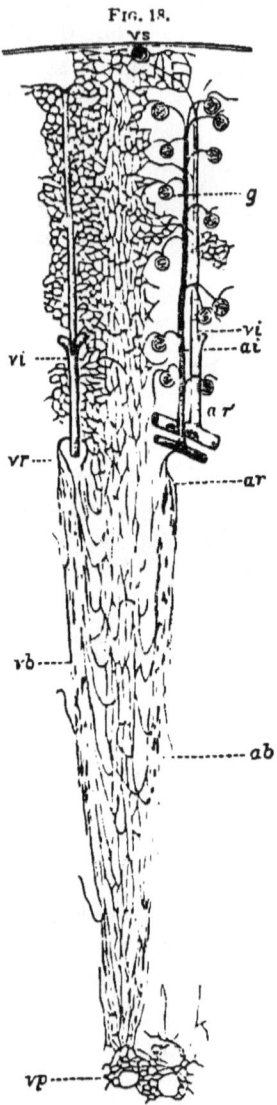

Diagram of the circulation in the kidney.—After LUDWIG.

before they reach the Malpighian capsule. A few only of the *vasa efferentia* lying adjacent to the medulla do not thus di-

vide, but distribute their branches in a manner presently to
be described. With these exceptions, the capillaries from all
the neighboring efferent vessels communicate with each other,
and the whole cortex becomes interpenetrated, as is well shown
in Figs. 16 and 18. The meshes of the resulting network surrounding the convoluted tubules are small and nearly round,
while those embracing the medullary rays are larger and wider.
According to Ludwig, the vessels are not closely applied to the
tubules, but lacuniform spaces, frequently filled with fluid, in-
tervene between the two. The terminal branches of the
interlobular arteries reach the capsule of the kidney.

That part of the network which surrounds the medullary rays
is continuous with the capillary system of the medulla, but the
blood from the cortical plexus is, for the most part, collected by
a system of veins (*ri*, Fig. 18) corresponding to the interlobular
arteries. Some of these, at least, begin on the surface of the
kidney by the union of a number of minute veins from the
external layers of the cortex, which unite in a radial manner,
so that the resulting vessels are called stellate veins (see Fig.
18, *vs*). Thence they descend, each vein in company with an
interlobular artery, towards the marginal layer of the medulla,
receiving numerous branches from the capillaries of the cortical
plexus above described. The interlobular veins which do not
begin in the stellate veins of the periphery begin in the labyrinth.

The interlobular arteries, with their afferent branches, and
the glomeruli may be compared to a bunch of ripe currants, the
interlobular vessel corresponding to the vertical stem of the
raceme, the afferent vessels to the lateral branches, while the
currant itself is the red glomerulus or capillary ball.

The Vasa Recta and the Bloodvessels of the Medulla.—It was
stated on p. 33 that not all the vasa efferentia split up into the
capillary system described, but that a certain number lying
adjacent to the medulla distribute their branches differently.
These branches are sent down directly into the medulla, in the
shape of straight vessels, comparable in their distribution to
the drooping branches of the willow-tree, and are called *vasa
recta*. But although straight *vessels*, they are *not arteries*, though
sometimes called *arteriolæ*, since they are without circular mus-

cular fibre-cells in their walls. They are simply elongated efferent vessels (See Fig. 18, *ar'*). These contribute, however, but a small proportion to the vasa recta. The majority, which are true arteries, are derived from the same arching branches which give off the interlobular vessels at the upper edge of the marginal layer of the medulla, but pass in the opposite direction. The straight vessels from these two sources, passing downwards, enter first the spaces between the bundles of tubules in the marginal layer of the medulla, dividing and subdividing in a brushlike manner (Fig. 18, *ar* and *ab*) until they finally become capillary, and embrace the tubules in wide meshes. The number of trunks of the straight vessels progressively diminishes as the tubules become closer and the papillæ are approached, until the latter are reached, when but one or two remain to form a plexus of capillaries on the papillæ themselves. The capillaries reunite to form veinlets, corresponding to the straight arterioles, and these veins accompany the arterioles in the fissures between the bundles of tubules, reuniting to form larger trunks, which finally empty into vessels at the border of the cortex, corresponding with those which give off the straight arterioles (see Fig. 18, *rb* and *vr*).

The system of the vasa recta is one of great importance in the circulation of the organ. It has been remarked that the capillary system at the border of the cortex is in communication with that of the medulla; but it is evident also that the vasa recta, derived from the same branches as the interlobular arteries, admit of the transit of a large amount of the blood of the kidney through a coarse-meshed capillary plexus without its passing through the complex system of the labyrinth. This is an important provision, which doubtless tends to preserve the integrity of the kidney in congestions, or when a large amount of blood is driven through it. Under these circumstances a short cut is, as it were, provided for the blood out of the organ, and the more delicate vessels of the labyrinth are protected from rupture.

We are indebted to Virchow and McDonnel for the discovery of the vasa recta, and to Virchow, Ludwig, and Beale for our present accurate knowledge of their arrangement.

III. *The Connective Tissue of the Kidney.*

From a pathological standpoint the connective tissue of the kidney becomes one of the most important of its elementary structures. All anatomists and renal pathologists, however, are not agreed upon its proportion and importance.

While none deny its presence in the medullary portion of the kidney there is still a difference of opinion as to its existence in the cortex. Goodsir, in 1842, described the connective tissue framework or matrix of the kidney in the cortex, as well as the medulla, as though it were as distinct from the vascular and tubular structures of the organ as the beams and rafters of a house are distinct from the materials between them. Von Wittich,* in Germany, first contested this view, claiming that the intervals between the tubules are occupied only by capillary vessels, and that there is no connective tissue in the cortex. In this he was for a time sustained by German histologists generally. Arnold Beer,† in 1859, reasserted the importance, from an anatomical point of view, of the connective tissue as a stroma or framework for the support of the tubules and vessels. Henle,‡ in 1864, asserted that demonstrable quantities of connective tissue are only found in the apices of the pyramids, where they occupy the intervals between the tubules. According to Henle it is a substance which in the fresh condition is clear and transparent, and it is only after the long-continued action of chromic acid and chromate of potash solutions that delicate fibres appear, containing round and elliptic nuclei. Further inwards, towards the cortex (boundary layer of the medulla), the nuclei become less numerous, and finally disappear altogether, leaving between the tubules only delicate layers of a finely fibrillated tissue, which cannot be considered an artificial product, because the fibrillæ are often ar-

---

* Von Wittich, Beiträge zur Anatomie der gesunden und Kranken Niere. Virchow's Archiv für Patholog. Anatomie, Bd. 3, 1849, s. 142.

† Arnold Beer, Die Bindesubstanz der menschlichen Niere im gesunden und krankhaften Zuständen. Berlin, 1859.

‡ Henle, J., Handbuch der Systematischen Anatomie des Menschen. Zweiter Band, zweite Lieferung. Harn und Geschlechtsaparat. Braunschweig, 1864.

ranged circularly around the tubules. In the cortex the
tubules and vessels are so closely intercalated that we can, as
a rule, only speak of a cementing (verbindenden) substance
rather than an intertubular and intervascular substance
(zwischen-substanz). " *True* connective tissue," says Henle,
" is found in the cortex and medulla only in the immediate
neighborhood of the bloodvessels."

At the present day Drs. Johnson and Beale, of England,
also assert that the connective tissue is of minimum importance, that it has no share in the anatomy of the organ, and
therefore, none in the pathology. Dr. Beale says :* " 1. In the
cortical portion of the kidney there is no evidence of the existence of a fibro-cellular matrix distinct from the walls of the
tubules and capillaries. 2. The fibrous appearance observed
in those sections of the kidney which have been immersed in
water is fallacious, and is due to a crumpled, creased, and col-
lapsed state of the membranous walls of the secreting-tubes
and capillary vessels. 3. A small quantity of transparent
material is to be demonstrated between the walls of the tubes
and the capillaries in health, and not even this can be detected at an early period of development."

Dr. Johnson says :† " The so-called matrix has no existence apart from the basement membrane and capillaries.
The convolutions of the tubes and the network of capillaries
mutually support each other. No connective or supporting
tissue is required ; and, as Dr. Beale well remarks, the inter-
vention of any such tissue would tend to increase the distance
between the secreting cells and the blood, and so render the
gland less fitted for the discharge of its function." . . . .
" If there be any connecting medium it is a homogeneous and
structureless element."

Modern histology has solved the problem consistently with
all observations. It is true that there is no distinct framework or " supporting " structure of connective tissue for the

---

* Beale, Kidney Diseases and Urinary Deposits. Philadelphia, 1869,
p. 23.

† George Johnson, Lectures on Bright's Disease. New York, 1874, p. 7.

tubules and vessels of the cortex, for they require none. At the same time, we know that no bloodvessel of any size penetrates a gland without carrying with it its adventitia, and although this diminishes with the growing fineness of the subdivisions of the artery, it is scarcely possible to draw the line where the cells of the connective tissue sheath entirely disappear. The interlobular arteries described are doubtless accompanied by such cells. Further, the researches of Ludwig, Kölliker, and Schweigger-Seidel have shown that the connective tissue of the cortex and of the marginal layer of the medulla is not of the fibrillar kind, but of the reticular or purely cellular variety; very delicate, soft, and scanty, as this kind of connective tissue usually is, winding its spindle and reticular cells in sparse numbers between the tubules and bloodvessels, yet sufficiently numerous to permit an isolated cell to be seen in almost any slide prepared from the fresh organ. These small isolated fusiform cells are described both by Ludwig and Schweigger-Seidel as being placed with their longer axes transverse to the direction of the convoluted tubules.

These two sources, therefore, the reticular connective tissue of the cortex and the adventitia of the bloodvessels, while they do not afford a matrix or framework, still furnish connective tissue cells in sufficient numbers to become the focus of an overgrowth. Further, in the case of the Malpighian bodies, no one who is in the habit of examining kidney sections under the microscope can have failed to see that they are surrounded by a capsule *of fibrillar* connective tissue, especially in the region of the medulla. This is a fact of great importance, because, in the case of diseased organs, it becomes necessary to distinguish such capsule from a pathological overgrowth of the same tissue.

As we recede from the boundary layer of the medulla towards the apices of the cones—the papillæ, the proportion of fibrous connective tissue increases until it becomes quite abundant in the latter region, surrounding the tubuli in a concentric manner.

It is even claimed by Axel Key that a reticular connective tissue formed of stellate cells is found uniting the capillaries and capillary lobules of which the glomerule is composed.

The Capsule of the Kidney.—The kidney is surrounded by a firm capsule, composed mostly of fibrillar connective tissue imperfectly laminated.   In health it is very loosely adherent by a looser and more delicate layer, to the cortex of the organ, whence it is therefore stripped with facility, the only attachments being a few filaments of connective tissue, and some small bloodvessels which pass from one to the other. In certain diseased states, especially where the connective tissue is involved, this adhesion is more intimate, so that it is impossible to strip off the capsule without dragging more or less of the secreting structure of the kidney with it; and in descriptions of post-mortem conditions, we constantly read that the capsule stripped off easily or was closely adherent, etc.   Eberth describes a plexus of unstriped muscular fibre-cells under the capsule of the kidney in man.

The blood-supply of the capsule is derived partly from those branches of the interlobular arteries which do not proceed as afferent vessels to the Malpighian bodies, and partly from branches of the phrenic, lumbar, and supra-renal arteries. These form a large-meshed capillary system, whence the blood is collected by corresponding veins, including the stellate veins of the cortex.

IV. *The Lymphatic Vessels of the Kidney.*

The extended information recently acquired with regard to the lymphatic system in general, only partially includes that of the kidneys, although the researches of Ludwig and Zawarykin* have furnished some facts of importance.   From them we learn : 1st, that the capsule of the kidney is provided with well-defined lymphatics possessing even valves; 2d, that similar larger trunks issue from the hilus of the organ along with the artery and vein ; and, 3d, that the spaces between the tubules

* C. Ludwig mit Zawarykin, Wiener Akademische Sitzungberichte, Bd. xlviii.

and vessels of the cortex are of the nature of lymphatic spaces. These spaces are less numerous in the medullary rays, and still less in the medulla. These observers have also shown that communication exists between the lymphatics of the capsule and those of the cortex, and between the latter and the lymphatics of the hilus, but the exact paths of such communication are not determined.

V. *The Nerves of the Kidney.*

While the nerves of the kidney doubtless play a most important rôle in its physiology, they are less demonstrably influenced by its pathological states than any other of the elementary tissues composing it. They are derived chiefly from the sympathetic system, and it is probably for this reason that, except in the case of acute Bright's disease and of impacted calculus, pain is a rare symptom in renal diseases, although this is contrary to popular opinion, which ascribes almost all pain in the back to kidney disease. We are indebted for many new facts, indeed for almost all we know about the minute distribution of the nerves in the kidney, to Professor Lionel S. Beale, M.B., of London. According to Dr. Beale, nerves, pale and dark-bordered (medullated), are distributed to the secreting-tubes and capillaries as well as the arteries ; connected with the nerve-bundles are numerous ganglia and ganglion-cells similar to those found in connection with the ramifications of the sympathetic generally in mammalia.* I have myself,† following Dr. Beale's methods, demonstrated these cells in connection with nerve fibres in the pig's kidney. Dr. Beale says further, that these " nuclei," so numerous on the nerves of the kidney, as well as on all the peripheral nerve-fibres of various tissues of mammalia, have been hitherto included among the connective tissue corpuscles of the matrix. He considers that the nerve-fibre distributed to the walls of the uriniferous tubules constitute an *afferent* system capable of influencing through the nerve-centres the *efferent* fibres distrib-

---

* Beale, Kidney Diseases and Urinary Deposits, 1869, p. 15.
† American Journal of the Medical Sciences, October, 1869.

uted to the arteries, the calibre of which they govern, and thus regulate the amount of blood flowing through the capillaries. The ganglia within the kidney are connected by nerve-fibres with those external to the organ, and with the spinal nerve-

Fig. 19.

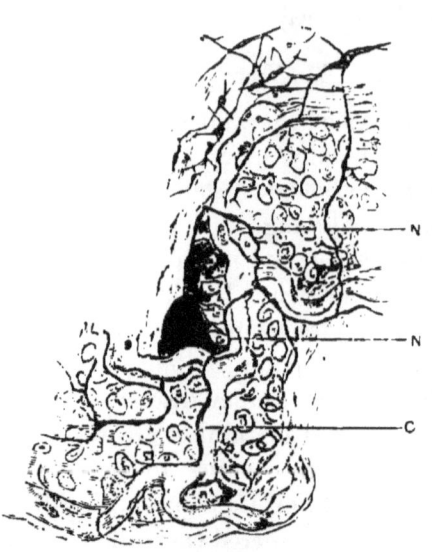

A part of a convoluted portion of a uriniferous tube from the newt's kidney, showing capillary vessels and nerve fibres, and the thickened basement membrane continuous in structure with the connective tissue. × 215. (After BEALE.) The finer dark lines, N, represent the nerve fibrils with nuclear thickenings. C, capillary bloodvessels.

fibres, and through these with the great nervous centres. And thus he would explain the well-known influence of the emotional centres over the secretion of urine.

VI. *Suggestions as to the Method of Studying the Histology of the Kidney.*

For a proper study of the minute changes in a kidney, whether healthy or diseased, it is necessary that sections should be made in two directions: 1st, longitudinally, or in the direction from the cortex towards the papillæ; 2d, transversely to this direction, or tangentially. In this manner the relation between the elementary tissues which make up the organ is preserved, and the changes in them may be studied understandingly. From cuts in the first direction one picture only is obtained, which may extend from the cortex to the papillæ. Cuts in the latter present a different picture, according as they are made through the cortex, or at different situations in a medullary cone—near the cortex, between it and the papilla, or at the papilla itself. It will be found that in the various forms of Bright's disease the cortex alone is generally involved, so that for the study of these conditions it is necessary only to make sections through

FIG. 20.

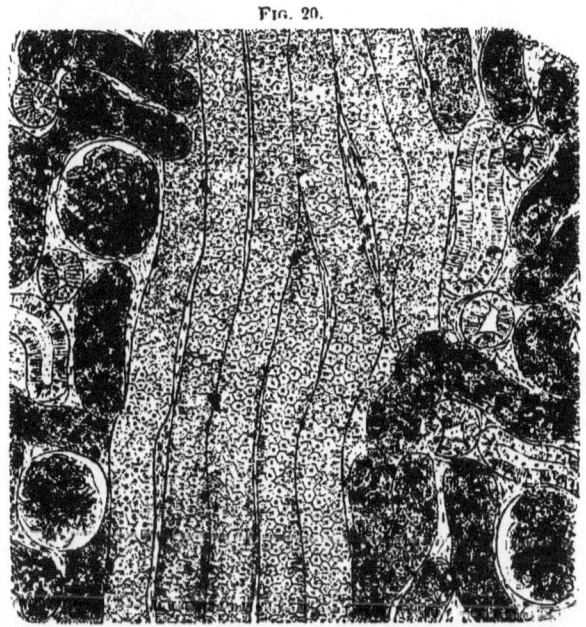

Longitudinal section through the cortex. Tubules forming medullary ray in centre, showing faintly outlines of epithelial cells. Malpighian bodies among sections of convoluted tubules; in several places these have been laid open, exposing the lumen of the tubule. The nuclei are seen imbedded in the dark granular protoplasm lining the tubules. The striated appearance of the cells is also well shown. Certain tubules are also transversely cut. × 160.

this portion. It is especially desirable that the normal picture should be familiar, in order that the abnormal may be understood.

Fig. 2 will give an idea of the appearance, under a low power, of a longitudinal section of an injected kidney. Fig.

6, however, presents the picture of a portion of the cortex
under a higher power, which is necessary for the correct appreciation of pathological changes.

In Fig. 3 is seen the appearance which will be afforded under
a low power by a transverse section through the cortex, a short
distance below the capsule, of the same kidney from which
Fig. 2 represents a longitudinal cut. It also affords a correct
notion of the lobular structure which the kidney really possesses, but which would not be suspected without the micro-

FIG. 21.

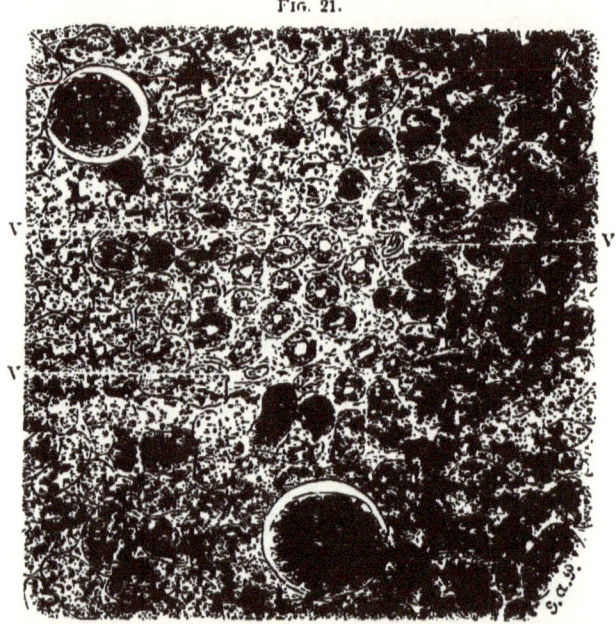

Tangential section through the cortex. Collecting tubes in centre cut transversely.
Two Malpighian bodies. Numerous sections of tubules, some in transverse section showing dark lumen of tubuli. Sections of several vessels, V, are seen near the central collecting
tubes. × 150.

scope. The renal lobules are seen to be polygonal figures, of
which the sides are formed by the branches of the interlobular vessels already described. The centre is occupied by the
straight tubes of the pyramids of Ferrein (medullary rays),
and between these and the border is the labyrinth, which is
the seat of the most important lesions of interstitial and paren-

chymatous nephritis. A section immediately at the surface, just under the capsule, would not give precisely the same picture, for it will be remembered that the medullary rays do not extend quite to the surface of the kidney. A section at this most superficial position would show only convoluted tubules and their intermediary portions. Not even Malpighian bodies would be here seen. In the longitudinal section, shown at Fig. 2, the brackets indicate the boundaries of the lobules. Figs. 20 and 21, show, under suitable enlargement, the appearance of transverse and longitudinal sections in the cortex. Fig. 22 a transverse section through the papillary portion of the medulla, showing the relatively large proportion of connective tissue between the collecting tubes in this situation.

To Prepare Sections of the Kidney for Minute Study.—As to the method of making sections for study, no means can

FIG 22.

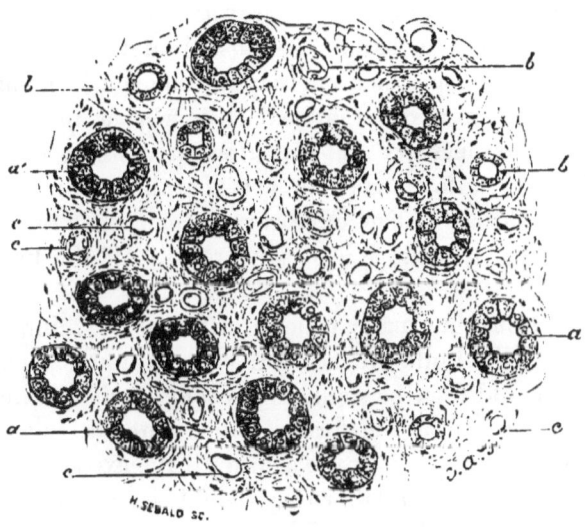

Transverse section through papillary portion of a medullary cone. *a.* Collecting tubes in transverse section separated by connective-tissue fibres and cells. *b.* Portions of loops. *c.* Bloodvessels. × 150.

produce cuts in which natural appearances are so well retained as the freezing microtom, which should always be used, if possible. But at present it is expensive and not always obtainable, and in its absence, portions of kidney may be hardened in alcohol, which should be gradually increased in strength from day to day; or Müller's fluid, which should always be changed after twenty-four hours and afterwards at longer intervals, until the tissue is thoroughly hardened. Sometimes, where haste is necessary, portions which have been for some days in Müller's fluid have to be transferred to alcohol before the hardening is sufficiently accomplished by the former; and of the two methods that by alcohol is always the more speedy. But this general truth may be laid down, that the more speedy the process of hardening employed, the more likely are natural appearances to be altered. The pieces of kidney thus hardened may be imbedded in a mixture of wax and oil or paraffin and benzin, or may be even compressed between suitable segments of carrot or turnip and cut with a razor of which one side is ground flat; or a section knife made expressly for the purpose may be used. The wax and oil mixture recommends itself over the paraffin, because from it there is no evaporation, and the specimen may lay imbedded a long time if it is not convenient to cut it immediately; whereas in the paraffin mixture it becomes dry and worthless after twenty-four hours. Paraffin and tallow make an excellent imbedding mixture, which has the advantage of always keeping close to the tissue imbedded in it, while it also slips readily off the knife after cutting.

The sections thus made, which are of desirable thinness when the edge of the knife can be clearly seen through them while cutting, should then be stained by a solution of carmine or of hæmatoxylon; if by the former, they should be immersed in a saturated solution of oxalic acid until the excess of carmine is removed; if by the latter they should be washed in distilled water. They are then returned to common alcohol. If it is simply desired to study them without preserving them permanently, they may then be clarified in oil of cloves, placed on a slide in a drop of the clove oil, covered with a thin glass cover and studied; or they may be transferred from the alcohol into glycerin instead of oil of cloves, and similarly mounted in a drop of the former. The glycerin also clarifies the section. Even thus mounted they may be kept for many days.

If the sections are to be permanently mounted, they are transferred from the oil of cloves into a drop of Canada balsam or dammar solution; the excess of oil of cloves being first removed by bringing the edge of the section in contact with a piece of bibulous paper. The balsam is best prepared by evaporating the ordinary Canada balsam to dryness and redissolving in chloroform, benzole or strong alcohol.

Sections may be also advantageously *doubly* stained by appropriate solutions of carmine and indigo-sulphate of sodium. The sections are first fully stained in the carmine, placed in acid solution, washed in water, and then transferred to a weak solution of the blue indigo-sulphate, where they remain until the satisfactory tint is obtained; again soaked in acid solution, and finally thoroughly washed in water, after which the usual process of preparation is followed. The double staining may be effected by a single immersion in a mixture of the two staining fluids, but the separate use of these is preferable, owing to the greater certainty of success and the facility with which the relative intensity of the tints is controlled.

VII. *Nature of the Act of Secretion of Urine.*

It is commonly known that two widely different views exist as to the nature of the act of secretion of urine. According to the much older view of Bowman, it is partly physical and partly vital. According to the later theory of Ludwig and his school it is a purely physical act. Ludwig claims that all the constituents of the urine are separated in very dilute solution by a pure act of filtration from the capillaries of the glomerulus into its capsule; and that in the convoluted tubules there occurs merely an aqueous reabsorption resulting in a concentration of the fluid. Professor Küss* goes so far as to

---

* Küss, A Course of Lectures on Physiology (second edition). Translated by Dr. Amory, Boston, 1875, pp. 465-467.

claim that the filtered product is really the *serum of the blood*, whence in health the *albumen*, as well as some of the water, is reabsorbed in the convoluted tubules; and Küss explicitly asserts that "the epithelium of the uriniferous tubes simply absorbs, but does not secrete."

According to Bowman, on the other hand, the filtrate separated in the Malpighian capsule is almost a pure water. In it may be dissolved some of the inorganic salts of the urine; but the most important ingredients, the organic nitrogenous,—including urea, uric acid, creatin, creatinin,—and extractives are added in the convoluted tubes.

It is now generally conceded that these organic constituents, as well as the inorganic, exist preformed in the blood; and although the experiments of Oppler,[*] Perls,[†] and Zalesky[‡] led to opposite conclusions, they lack confirmation, and now few, if any, claim that the urea and extractives are manufactured, so to speak, in the kidney. They are believed to exist preformed in the blood, and are selected therefrom by the cells of the convoluted tubules.

A priori there are two important reasons why we should expect that the proper gland-cells of the kidney should have some more highly specialized function than that of aqueous reabsorption: 1st. The fact that they possess the anatomical peculiarities of glandular epithelium whose function everywhere else in the economy is an elaborating one. 2d. The important reason, drawn from pathology, that when, as the result of disease, this epithelium is practically stripped from the tubules we have the retention of urea and often the phenomena of uræmia. But notwithstanding these important facts, the tendency of modern physiology has been to accept the purely physical view of Ludwig until the researches of R. Heidenhain were published. The indigo sulphate of sodium (indigo-

---

* Oppler, Beiträge zur Lehre d. Urämie, Virchow's Archiv, vol. xxi, p. 260.

† Perls, Beiträge zur Lehre d. Urämie, Konigsberg Med. Jahrb., vol. iv, p. 56.

‡ Zalesky, Untersuch. über d. Urämischen Process, Tubingen, 1865.

carmine) is a substance which, when injected into the blood of
animals, is promptly separated by the kidneys and appears in
the urine. Heidenhain injected this substance, and after appropriate intervals, removed the kidneys and examined them
microscopically. In no instance did he find any of the indigo-
carmine in the Malpighian capsules; but the cells of the convoluted tubules and the ascending loop of Henle were filled
with it, as was also the lumen of the tubes if the animal was
killed sufficiently long after the injection.

Similar experiments with urate of sodium showed that this
substance is secreted in precisely the same situation.* With
these facts before us it no longer seems reasonable to deny a
direct selective action on the part of the cells of the convoluted
tubules, and I accept, therefore, as the correct view of the
nature of the urinary secretion, that the water of the urine is
filtered out in the Malpighian capsule, the condition favorable
to such filtration being supplied by the increased blood-pressure
which exists in the glomerulus. In this water may be dissolved some of the inorganic constituents of the urine, but the
most important nitrogenous principles, the true effete and
poisonous matters which it is the office of the kidney to remove,
are separated by the agency of the cloudy cells lining the convoluted tubules, the ascending limb of Henle's loop, and the
intermediary segment of Schweigger-Seidel. From these cells
they are pushed out into the lumen of the tube by the *vis a
tergo* of additional secretion, and dissolved by the water which
comes down from the Malpighian capsule, thus producing the
urine, which is gathered up by the collecting and excreting
tubes, by which it is emptied into the pelvis of the kidney at
the papillæ.

* The details of these experiments are very interesting, but the reader
must be referred to the paper of J. Heidenhain in Max Schultze's Archiv,
vol. x, 1874, p. 1; also to his paper, Versuche über den Vorgang der Harn-
absonderung, in Pflüger's Archiv, vol. ix, p. 1, 1874. Good abstracts are
also found in Foster's Physiology, p. 277 (first edition), 1877, and in Charcot's Treatise on Bright's Disease, translated by Millard, 1878.

SECTION II.

TESTING FOR ALBUMEN.—ALBUMINURIA, ITS SOURCES AND MECHANISM OF ITS PRODUCTION.

TESTING for albumen lies at the bottom of all diagnosis in Bright's disease, and there is no form of kidney affection in the investigation of which we dare omit to determine the presence or absence of albumen. Further, I am confident that even at the present day small amounts of albumen are not unfrequently overlooked in hasty examinations. Hence it cannot be out of place to discuss briefly this matter.

When large quantities of albumen are present in urine, its detection is most easy. The addition of from 5 to 30 drops of acid—nitric is commonly used—to half an ounce of urine in a test-tube, will promptly precipitate the albumen, if much is present. Or, if the urine is acid in its reaction, the albumen is as promptly thrown down by boiling; and if, after boiling, a few drops of acid are added, the solution of any phosphates precipitated by the heat is secured; while any albumen which may have remained in solution because the original urine was alkaline when boiled, will be thrown down. It should not be overlooked that, unless a considerable amount is present, albumen is not precipitated by heat from an alkaline urine.

But it is the small amounts of albumen which escape attention, and it is to these that I would especially direct attention. Under ordinary circumstances, by far the most distinctive test for small quantities of albumen is that form of the nitric acid test known as Heller's, to be presently described. But, in the course of an experience involving almost daily examinations of urine, I have met a number of instances in which it failed to give evidence of the presence of albumen, when the ordinary heat and acid test applied in the manner to be described, proved it conclusively.

Many who have often tested urine for albumen by the ordinary heat and acid test, will have observed that after boiling the clear urine, and adding a few drops of nitric acid, the resulting fluid will be apparently clear ; but upon setting aside the urine thus treated, say for twelve hours, or until the next morning, there will sometimes be found a small deposit. Supposing the urine before testing to have been *carefully filtered*, this deposit is either : 1st, acid urates ; 2d, uric acid; 3d, nitrate of urea ; or, 4th, albumen. The *first* result from a partial decomposition of the neutral urates by the nitric acid added ; the *second* by a further action of the acid upon the acid urates, and a resulting complete separation of the uric acid from the sodium, potassium, etc., with which it was combined ; the *third* is found only when the urine happens to be highly concentrated and contains an unusual proportion of urea. The *second* and *third* have well-known forms of crystallization, by which they can be easily recognized under the microscope, but the acid urates and albumen are both amorphous and cannot therefore be thus distinguished. *All*, however, *except albumen*, disappear on the reapplication of heat. In all doubtful instances, urine which has been tried by heat and nitric acid *should be boiled again* after cooling and standing from six to twelve hours, and if the sediment is not dissolved by such ebullition it is albumen.

My own method, therefore, of examining a specimen of urine for albumen is invariably as follows :

I. Unless *perfectly* clear it is first filtered. A portion of the filtered urine is then *boiled*, and carefully watched in a *suitable light* for detection of the least diminution in transparency.* I say a *suitable* light, because it is not always a bright

---

* In these directions I have omitted the one of previous acidulation contained in the similar directions in my little book on the Practical Examination of Urine, third edition, 1880, p. 33. I do this because, in the first place, in a large number of instances, the urine is acid to begin with, and in the second, the addition of the drop or two of acid *after* boiling accomplishes the same purpose, while it also dissolves any phosphates which may have been precipitated by the heat. On the other hand, the previous acidulation does not obviate the necessity of the addition of the drop of acid after boiling, and in the method given in the text, one manipulation, and, therefore, time, is saved.

light that is desired. A diminished transparency might easily
be overlooked in it, and it is generally desirable to shade the
light a little with the hand or a book before the diminished
transparency can be noticed, if the latter is slight. On the
other hand, an insufficient light, such as is shed on a dark day,
is altogether unsatisfactory. A drop or two of nitric acid is
then added, and if any turbidity which has resulted from the
action of the heat, disappears, such turbidity is due to phosphates and not to albumen. It is sometimes desirable to compare the tube and contained urine with another filled to the
same depth with urine which has not been heated. If any
degree of turbidity remains, it is caused by albumen, and the test
may end here, although it is well to put the test-tube aside, in
order that, after subsidence, the albumen may be approximately
estimated. If, however, there is the least doubt about the
presence of albumen, the tube must be set away, carefully protected from dust, for from six to twelve hours, in order that
any appreciable sediment may subside, and be subsequently
again tried with heat.

II. Another small test-tube is then filled to the depth of
half an inch with colorless nitric acid. About as much urine
is then allowed to fall gently upon it from a pipette, in the
manner represented in the figure, while the point of junction
of the two fluids is carefully watched for the white line of albumen. This is best seen by shading the tube by the hand or
a pamphlet placed behind and slightly below the line of junction, so that the light may fall obliquely upon it. The gradual
delivery of the urine from the tube is greatly facilitated by ro-
tating the latter under the index finger, especially if the upper
end of the tube is a little roughened. This secures such a
gradual entrance of the air as permits the urine to descend
very slowly without mixing with the acid below.

Precautions.—In urines of high specific gravity, from concentration of the ordinary constituents, a white band makes
its appearance at the border between the two fluids, which
is not albumen, but *acid urates*, resulting from the partial
decomposition of the neutral urates and the precipitation of

the more insoluble acid salt. But the behavior of the band thus produced is quite different from that of the albumen.

F.g. 23.

Testing for albumen by nitric acid.

Although sharply defined at first, the lower border stands at a slightly higher level than that of the albumen, while the upper edge soon ceases to be circumscribed and rises upward into the urine like a cloud, soon pervading the entire supernatant urine; whereas the albumen remains sharply defined at the line between the two fluids, until, in the course of several hours, if its quantity be small, it may be completely dissolved by the excess of acid present. Further, the cloud of acid urates is promptly dissolved by the application of a moderate heat. If, as sometimes happens in acute fevers, the urine is both concentrated and contains albumen, there may be two layers, an upper cloudy one of urates, and a lower sharply defined of albumen. But in chronic Bright's disease, at least, urine is generally of low specific gravity, and the complication is not likely to occur.

When these precautions are observed in connection with the above double test, it is almost impossible to err as to the presence of albumen.

It sometimes happens, however, that, in consequence of age and putrescence, the urine has become alkaline, precipitated phosphates pervade the fluid, and it swarms with bacteria. As a consequence, it is so opaque that small quantities of albumen cannot be detected. Filtering does not mend this. The bacteria and finely divided phosphates pass through the minute openings of the filter with the filtrate. Under these circumstances it becomes necessary to clarify the urine. For this purpose I use a method suggested by Hoffmann and Ultzmann. Boil a portion of the urine with an *equal* bulk of officinal liquor potassæ, and filter. If still not quite clear, add a few drops of the "magnesian fluid,"* warm again, and filter. To the clear fluid thus obtained Heller's acid test may be applied, or, after careful acidulation, the *heat* test. In this manner very small quantities of albumen may be recognized in urine which would not otherwise permit its detection.

To Indicate the Proportion of Albumen in a given Specimen of Urine.—In practice it is scarcely possible to apply the only exact quantitative method of estimating albumen, by coagulation, filtration, drying, and weighing, because of the time it necessarily occupies. Even the method suggested by Dr. William Roberts,† of Manchester, is for the same reason unavailable to the busy practitioner. Nor can it be said that these more accurate methods are necessary. For practical purposes it is sufficient to compare from day to day the bulk of precipitated albumen with that of the urine tested. For this purpose it is convenient to have a test-tube graduated to one-half its depth, so that after coagulation by heat and

---

* The *magnesian fluid* is made by dissolving magnesium sulphate and pure ammonium chloride, each *one* part, in *eight* parts of distilled water, and adding one part of liquor ammoniæ.

† American Journal of the Medical Sciences, January, 1878, p. 209, from the Medico-Chirurg. Transactions, vol. xli, 1876. Also, the 2d ed. (1878), p. 39, of the writer's work on The Practical Examination of Urine. For the reason mentioned in the text it was omitted from the 3d ed.

acid, and allowing sufficient time for subsidence, the estimation can be made. In writing or speaking, this should
be referred to as one-half, one-fourth, or one-tenth the bulk
of urine tested, as it may happen to be. A very careless
habit which exists of speaking of these proportions as *per-
centages* should be avoided. Thus, when the albumen in a
given specimen of urine equals one-fourth its bulk, it is often
carelessly said that there is 25 per cent. of albumen. Now when
we remember that the serum of the blood does not contain more
than 5 per cent. of albumen, and the most highly charged
albuminous urines rarely contain more than 2 per cent., the
absurdity of this mode of expression is evident. Only when
the albumen is coagulated, dried, and weighed, and the weight
compared with that of the urine whence it is taken, should we
speak of percentage; otherwise the proportion of bulk is referred to, and it should be clearly indicated.

*Seat of Transudation of Albumen in the Albuminuria of Renal
Disease.*—It is commonly believed that when urine is albuminous, the albumen transudes along with the water into
the Malpighian capsule, although there has never been any
absolute demonstration that this is the case. Experiments
of Heidenhain, with a view to settling the question, led to
negative results, and until some positive demonstration is
afforded to the contrary we may accept on theoretical grounds
the view usually held. These lie in the fact that in the first
place there is a physiological constriction at the orifice of the
efferent vessel by which the blood leaves the glomerulus. This
is of course increased under pathological conditions when there
is hyperæmia of the second capillary network, producing swelling and overgrowth of the cells of the convoluted tubules,
which compress the bloodvessels of the rete, and further resist
the egress of the blood from the glomerulus. Under these
circumstances we have the increased pressure under which
albumen will transude, although its osmotic equivalent is very
low, and under normal conditions it will not pass out. This
well-known property of albumen is a sufficient reason against
the view of Küss, already alluded to, page 47, that under

normal conditions the transudate into the Malpighian capsule
is the serum of the blood, from which the albumen is reabsorbed by the cells of the convoluted tubules.   It is hardly
necessary to say that this view, if correct, would explain an
albuminuria which might be present either when the tubules
are stripped of their cells or the latter are functionally inert
from degeneration or malnutrition; and indeed Küss makes
use of this fact as an argument in its favor.

*Albuminuria from other Sources than the Parenchyma of the
Kidney.*—It is well known that albumen may enter the urine
from other sources than the secreting substance of the kidney.
The pelvis of the organ, the ureters, the bladder, and the urethra, and in the female, the vagina and uterus in addition, are
these sources.   In all of them except the uterus it is almost
invariably the serum of pus formed during inflammation of
their mucous surfaces which furnishes the albumen.   The
presence of pus-corpuscles, therefore, in sufficient number in
the urine is usually sufficient to explain the source of such
albumen, which is moreover usually small in quantity.   It
must not be overlooked, however, that the two sources, that
of the kidney itself and the mucous surfaces, referred to, may
be combined, in which event careful microscopic examination would sooner or later discover tube-casts, while the quantity of albumen would be larger than could be accounted for
by the presence of pus alone.

I have, however, in several instances of disease of the bladder where there had been suppuration and albuminuria from
such disease, found that with an insignificant residue of pus
there remained an albuminuria, not large, but altogether dis-
proportionate to the amount of pus remaining, and where the
most careful examination failed to discover renal disease.
These cases I cannot satisfactorily explain, and allude to them
here hoping the attention of others may be called to the matter.

Bartels* mentions copious albuminuria, with an inconsiderable sediment of pus, after the free use of cantharides internally

---

* Bartels, Carl, article on Diseases of the Kidney, in Ziemssen's Cyclopædia of Medicine, vol. xv, p. 31, New York, 1877.

and externally, as a form of albuminuria to be distinguished from that which so constantly accompanies Bright's disease; but it seems to me the condition thus resulting ought to be regarded as the first stage of acute Bright's disease or parenchymatous nephritis.

Menstrual or lochial blood need only be referred to as sources of albumen in the urine, hardly likely to be overlooked by any physician; while hæmorrhage from any one of the mucous surfaces referred to, as well as the kidney itself, would be a source of albuminuria.

It is usually comparatively easy to determine whether a hæmorrhage has its source in the kidney on the one hand or the mucous membranes above mentioned on the other. In the former, coagula are never present, but the blood entering the pelvis of the kidney, slowly and intimately mixed with the urine, imparts to it when acid in reaction a *smoky hue* which is very characteristic. The coloring matter of the corpuscles is mostly dissolved out by the urine which it thus tinges, and on standing the stroma of the corpuscles sinks to the bottom in the shape of a brownish sediment. The microscope reveals these corpuscles shrunken, almost colorless, and often crenated. I have said that the smoky hue is present only in acid urine. When the latter is alkalized, either by spontaneous or artificial change in reaction, it assumes a brighter red hue, the degree of which depends upon the quantity of blood. The same cause, acidity, produces the smoky hue of blood which is vomited, and therefore mixed with gastric juice. When blood comes from the pelvis of the kidney or the ureter in any quantity, coagula, which are moulds of the ureter, are sometimes found, the descent of which is often attended with severe pain.

Another source of albuminous urine, though not likely to cause error, should be mentioned, and that is the so-called chylous urine, or chyluria, in which, in consequence of some as yet imperfectly understood communication between the lymphatic system and the urinary tracts, chyle enters the urine and imparts its physical and chemical characters more or less thereto. These are a milk-white appearance, due to the presence of fat in a molecular state of division, and a large amount of albumen. The disease is one of tropical countries usually, but I have met two cases in this latitude.

The Immediate Cause of the Transudation of Renal Albumen.—It has already been intimated that obstruction to the onward movement of the blood is the immediate agency which forces albumen, an otherwise non-osmotic substance, through the walls of the capillary bloodvessels. It was at one time claimed that a peculiar state of the albumen of the blood in disease facilitated its transit, but the exhaustive experiments of Stockvis* have refuted this view, which has no longer any supporters. The required obstruction is produced by any cause which sufficiently resists the movement of the blood through the kidneys, whether that cause resides in the organ itself or the venous system beyond it, whence the albuminuria which so often attends extreme valvular disease of the heart.

It is equally certain, however, that the transudation is facilitated by changes in the capillary walls. In the first place it is conceivable that some change in the texture of the membranes takes place analogous to such as occurs in a filtering membrane through which a fluid is forced by pressure, an increase in the size of the apertures or pores of the membrane if we so choose to term it. But it must be remembered that this is a direct result of the intravascular pressure and would not exist without it. On the other hand, the experiments of Cohnheim upon inflammation† have shown that among the phenomena of inflammation a free transudation of albumen accompanies the wandering out of the colorless corpuscles of the blood. This element of increased albumen exudation would be found only in the more acute inflammatory conditions of the kidney, and accounts for the increased quantities of albumen in these as compared with the more chronic processes where pressure is the sole agent.

---

* Stockvis, Recherches Expérimentales sur les Conditions Pathogéniques de l'Albuminurie, in Journal de Médecine de Bruxelle. Vols. 44, 45, 1867.

† Cohnheim, Ueber Entzündung, Virchow's Archiv, Bd. 40, S. 77.

The effusion of albumen takes place, not only from the glom-erulus into the uriniferous tubules of the kidney, but also in the interstices of the organ outside of the tubules, wherever the colorless corpuscles wander, furnishing there a pabulum for the overgrowth of the intertubular and intervascular connective tissue, which always takes place if the process be sufficiently long continued.

It is possible, also, that small quantities of albumen pass into the uriniferous tubules of the kidney, which would not otherwise do so, in consequence of alterations in the epithelial lining of the tubules, the result of disease. But the quantity from this source must be very small, and the fact remains that intravascular pressure, aided by alterations in the capillary walls in inflammation, is the chief immediate cause of albu-minuria, at least in acute renal disease.

In chronic albuminuria another important influence operates in facilitating the transudation of albumen. This is the hy-drœmic state of the blood, which is itself a consequence of albuminuria.

One other source of renal albumen requires to be alluded to, although not really a symptom of kidney disease as ordi-narily understood, that is, it is not due to structural alteration of the organ. It is the albumen which appears in the urine in grave cases of infectious disease, as typhoid fever, small-pox, etc., and even in local inflammations other than renal, attended with much fever, as pneumonitis. That albuminuria of diphtheritis and scarlatina, which is due to an intercurrent parenchymatous nephritis, is of course not referred to. The albuminuria alluded to is not usually large, and disappears with the decline of the disease. Gerhardt* first suggested that this albumen represented the disintegration of numerous red blood-disks, and the observations of Obermüller† tend to the same results. But Bartels thinks it unnecessary to adduce such an

---

* Gerhardt, Ueber die Eiweissstoffe des Harns, in Deutsches Archiv für Klin. Med., Bd. 5, S. 213.

† Obermüller, Beiträge zur Chemie des Eiweissharns. Inaug. Dissertat. Würzburg, 1873.

explanation, on the ground that globulin, the albumen of the blood-corpuscles, possesses different properties from *serum albumen*, which is the albumen of urine. He agrees with Gerhardt that in subjects of acute fever, whose temperature stands persistently above 40° Cent. (104° F.), albumen passes into the urine, because the process of filtration by which the urinary secretion is effected in the Malpighian bodies, is, by reason of the abnormally elevated temperature, conducted under entirely abnormal conditions; that the elevated temperature relaxes the walls of the bloodvessels, or makes them yield more than they otherwise would, to the hydrostatic pressure of the blood streaming through them. But it is the nature of the filter, by which the urine is secreted, that is disarranged; for, inasmuch as the surface of the filter is enlarged, its pores must of necessity become more roomy.* Finally, he concludes that febrile albuminuria is akin to the overflow of albumen into the urine which follows section of the vasomotor nerves of the kidney, the excessive heat of the body acting the same way as such section.

This seems a circuitous method of accounting for what may be explained in a much simpler manner, although of course theoretically also. It is to be remembered that it is only in the cases of most intense fever, where the temperature is persistently above 40° C. (104° F.), that albuminuria, independent of renal disease, presents itself; also that in these there is the most rapid destruction or combustion of tissue of all kinds, including the corpuscles of the blood; that in the course of such combustion there must inevitably be incomplete oxidation of its products, and that such products would naturally find their way into the urine in the shape of albumen, just as among the ashes of a rapid and imperfect combustion in a furnace, portions of partly burnt fuel are found. The exudation of such albumen would be facilitated by the tendency to stasis which exists in the kidney in common with all organs, but is greater in it in consequence of the peculiarities

---

* Bartels, op. cit., p. 53.

of its capillary circulation. Such albumen might be supposed
to have greater diffusive power than belongs to typical serum
albumen, since, if time and favorable circumstances permitted,
it would have been entirely converted into urea, and, as such,
removed through the kidneys. It would be comparable rather
to the extractives, which are also partially oxidized products,
which pass from the blood into the urine.

SECTION III.

CASTS OF THE URINIFEROUS TUBULES—THEIR NATURE AND CLINICAL SIGNIFICANCE.

A so-called "urinary cast" or "tube-cast" is a mould of a uriniferous tubule. To this all are agreed. But beyond this, opinions differ not a little as to its exact mode of origin, composition, and significance. I retain the term "cast," in common with most English physicians, in preference to that of "cylinder," because it conveys the idea of what is meant with sufficient preciseness, while it implies nothing more. The adjective "fibrinous" I also discard as conveying an erroneous impression as to the composition of many tube-casts, although it is true that some are fibrinous.

Casts may be studied as a sediment in the urine or *in situ* in the uriniferous tubules where they are formed. They are characterized by differences in appearance and structure, as the result of which they are variously named. Thus there are hyaline casts, epithelial casts, blood-casts, oil-casts, waxy casts, and granular casts. Of the latter there are two subdivisions,—the *pale*, or moderately granular, and the *dark* granular casts. As to the material of which they are composed different views are held. I will first present my own, because it will enable me at the same time and in the briefest manner to describe the several varieties.

In the first place, all casts are not of the same composition. Thus, what are known as *blood-casts* are made up of coagulated fibrin and blood-corpuscles, the latter being entangled in the coagulum precisely as they are in a clot formed out of the body. The blood, by transudation or because of rupture of the capillaries of the Malpighian tuft, trickles down into the tubule, where it coagulates, forming a mould of the tube, and, after contraction, slips out into the pelvis of the kidney, whence it passes by the usual channel into the bladder. Blood-casts are represented in Fig. 24.

FIG. 24.

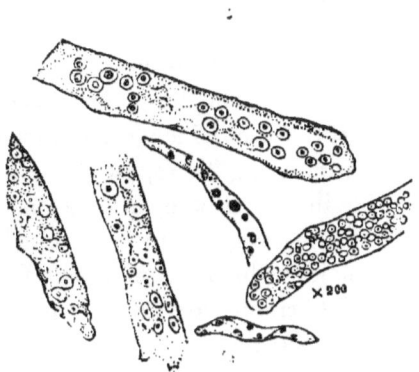

Blood-casts.—After WHITAKER.

Second, the term *epithelial cast* (Fig. 25) is applied to any cast to which epithelial cells are attached, be they few or many. In the latter case, the entire mould is often made up of closely-

FIG. 25.

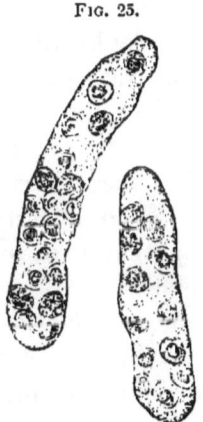

Epithelial casts. × 200.

packed desquamated cells, which may be cemented by their own adhesive properties or by a small quantity of fibrin, from simultaneously effused blood ; for blood-casts and epithelial casts constantly accompany each other. When the epithelial cells and their fragments—often their nuclei only—are few, they are separated by intervals, at times quite regular, of a hyaline material, to which they may be attached, or in which they may be imbedded. And it is the nature of this hyaline material which is the subject of dispute. It seems most reasonable to consider it an exudation from the blood of a fibrinous or an albuminoid substance, which coagulates after entrance into the tubules, entangling there, as does fibrin, whatever substances may be in the tubule, and occluding it for the time being. That it is true fibrin coagulated is not inconsistent with the modern doctrine of fibrin formation. For this, fibrinogen and fibrinoplastin are essential. The former is furnished by the blood-plasma, the latter by the blood-corpuscles, and as the latter are constantly present, although it may be not always in large numbers, the conditions of coagulation are furnished. Where there is haemorrhage into the tubules, both conditions of coagulation are abundantly present and blood-casts are also formed. It is true, the chemical properties of casts have been investigated, by Axel Key, and also by Rovida, who assert them to be altogether different from those of fibrin. But, that the hyaline material is in many instances some sort of exudate rather than a result of a metamorphosis of cells, is sustained by the fact that the hyaline is the only shape of cast formed in the slightest derangements of the kidney, as when there is a mere transitory congestion of the organ, and where there is no alteration of the epithelium whatever. The same objection holds to their being a secretion of the cells lining the tubules.

Third, the two forms of *granular cast* have probably different modes of origin. In the case of the *dark granular cast* the cells have undergone complete disintegration, and the products of this form a closely packed, dark, granular mass. If many red blood-disks happen to be in the tubule and share the breaking up, a yellowish tinge is communicated to this form of cast. In other instances the granular matter is less abundant, and a *pale* granular cast is the result. The granules here are less easily accounted for. They may result from a similar disintegration of cells fewer in number, or they may be due to

Fig. 26.

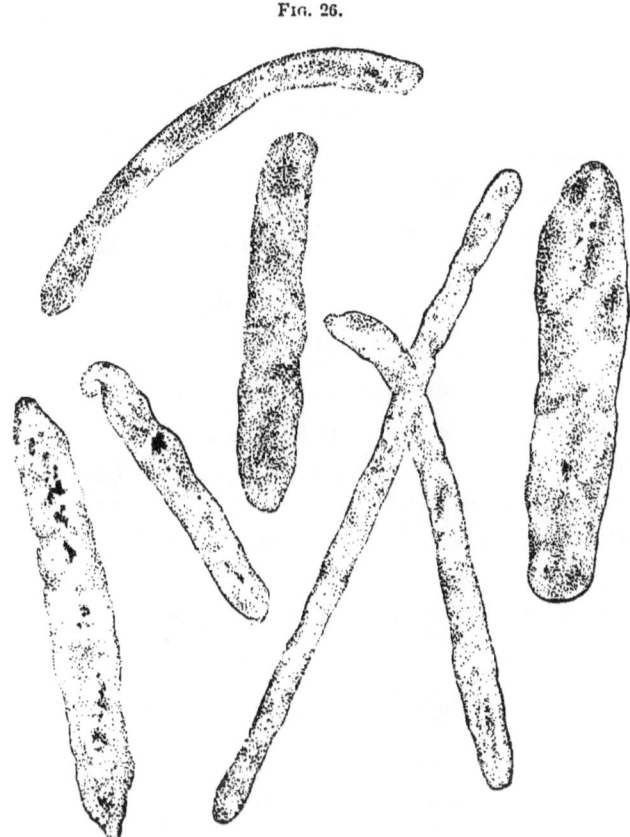

Granular casts. × 225.

a precipitation of granules from the albuminous substance of the hyaline cast, similar to that which occurs in the so-called cloudy swelling of cells. They would then be the result of the further metamorphosis of a hyaline cast.

Fourth, the *oil-cast* is a like product. Here the cells have undergone a fatty degeneration, some having completely broken up into fatty globules, while other cells retain their continuity

5

and shape, forming compound granule-cells and other forms of
fatty cells.   The oil-drops and fatty cells are seldom so abun-

FIG. 27.

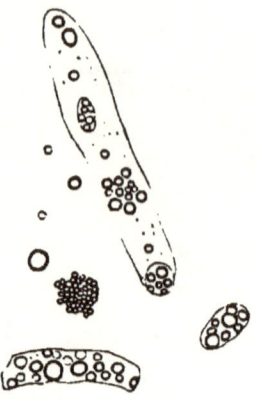

Oil casts and fatty epithelium.

dant that they form a closely packed mass constituting the cast,
but they are more or less separated by the hyaline basis referred to.   Fatty casts in urine are apt to be accompanied by
free oil-drops and free fatty cells, as seen in Fig. 27.

FIG. 28.

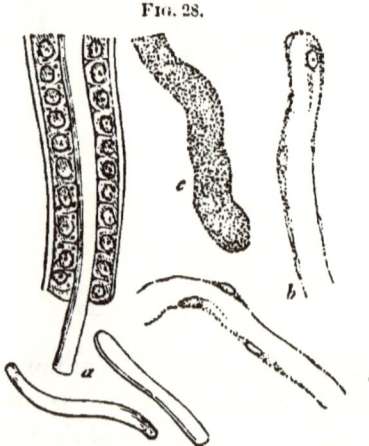

Pale granular and hyaline casts, one of the latter of small diameter, protruding from
a tubule in which the epithelium remains intact. A cast of the same tubule bereft of
its epithelium would evidently be of larger diameter.—After RINDFLEISCH.

Fifth, *hyaline casts* are the simplest form of cast.   They are
delicate, structureless, or almost so, and composed of the transparent substance already referred to.   They are of different

width, even when produced in the same tube, according as the cast is formed when the tubule is bereft of its epithelium or

FIG. 29.

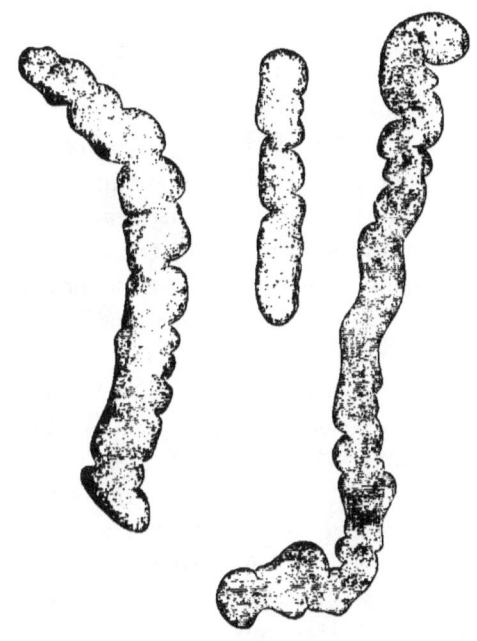

Hyaline casts. × 225.

the latter remains adherent.   This will be understood by examining Fig. 28.   In the former case a hyaline cast of "large" diameter is produced, in the latter one of "small" diameter. The range of width thus produced is from .025 mm. to .05 mm. ($\frac{1}{1000}$ to $\frac{1}{500}$ of an inch).   Hyaline casts are sometimes so delicate and transparent as to be scarcely visible, and require care in illuminating the field of view of the microscope to bring them out distinctly.   They are also often characterized by peculiar indentations or partial fractures, as shown in Fig. 29. Hyaline casts occur in all forms of renal disease.

Sixth, *waxy casts* are also hyaline, but more solid and glistening in appearance, more highly refracting, and apt to present a yellowish tinge.   From their supposed resemblance to molten wax, they have been called "waxy," but they bear no essential relation to the so-called "waxy" or lardaceous or amyloid kidney, by all of which terms it is known. The material of which they are composed has, at least, different refracting properties from that of the ordinary hyaline cast, and probably originates differently. It will be seen, presently, that it is this form of cast with regard to which there is most unanimity in ascribing it to a fusion and hyaline transformation of desquamated epithelium, or of other cells within the tubules. The same objections which lie in the way of admitting such an origin of the simple hyaline casts do not exist here, and it is not unlikely that they thus arise. It should be clearly understood, however, that the transformation is not the so-called waxy or amyloid transformation, the presence of which in the bloodvessels, tubules, and cells of the kidney constitutes the form of kidney disease known under that name. For these waxy casts do not, as a rule, strike the iodine reaction of the amyloid change. That they do occasionally, is attested by the dark mahogany-red tint of the two casts pictured in the colored plate opposite, a result I obtained, however, only after many times carefully and systematically treating these casts with iodine. The more usual coloration by iodine is that of the yellow casts shown in the same plate. Bartels* has obtained the same reaction in two instances, and suggests that the age of the material forming the cast may have something to do with the transformation into true amyloid material. And Rindfleisch† also says that he is convinced that casts which are long retained in the urinary tubules, especially in the bends of the looped tubes, undergo a glassy swelling, and assume the microchemical characters of amyloid substance, striking the iodine reaction, even in kidneys which are not otherwise amyloid. That they are not always the altered epithelium of the tubes is evident, for the reason assigned by Rindfleisch; in the first place the epithelium is seen well preserved between the casts

---

* Bartels, in Ziemssen's Cyclopædia of Medicine, vol. xv, p. 90.

† Rindfleisch, Patholog. Histol., New Syd. Soc. Transl., vol. ii, 1872, p. 144. Fifth German edition, 1878, p. 443.

and the basement membrane, and in the second, it is only in the highest degrees of the amyloid degeneration of the kidney that the cells are involved.

Again, waxy casts are found in all forms of chronic renal disease, it is asserted by some, in chronic interstitial nephritis more frequently than in the amyloid form. But I have seen them very numerous, also, *in situ* in the large white kidney.

Views as to the Nature and Formation of Tube-casts.

Dr. A. Burkhart, in a successful prize essay* presented to the medical faculty of Tübingen, in 1871, furnishes an interesting historical sketch of the study of casts, from the time of their discovery to the date of his essay. Although Simon is the reputed discoverer of casts, several observers had previously published descriptions of objects found in urine, which were evidently casts. The first of these was Vigla, who, in a paper on the microscopic constituents of urine, published in a journal called *l'Expérience*, in 1837, spoke of "finely granular longish albuminous plates." He gave no further description, and did not state in what kind of urine they were found. In 1838 Donné, in the same journal, denied the occurrence of these albuminous plates. To this Vigla replied that he had observed them in the urine of Bright's disease. In the same year, 1838, Rayer, in his treatise on diseases of the kidneys, described certain membranous, irregularly shaped, finely granular, white or yellow lamellæ, as occurring in albuminous urine, but said nothing of their composition. The first unmistakable description of these bodies was given by Nasse, of Marburg, in a notice published in Schmidt's *Jahrbuch*, in 1838, of Donné's plates of urinary sediments, and more clearly in a paper by him on the microscopical constituents of the urine in Bright's disease, published in 1843.† Simon first

* Burkhart, A., Die Harncylinder mit besonderer Berücksichtigung ihrer diagnostischen Bedeutung. Gekrönte Preisschrift. Mit Einer Tafel. Berlin, 1874.

† Nasse, Medizinischen Correspondenzblatt rhenischer und westphalischer Aerzte, No. 8, s. 121, 1843.

described casts as constituents of urine in Bright's disease in his work on practical chemistry, in 1842.

Henle was the first to study casts *in situ* after he had previously examined them in the urine of a case of Bright's disease. This he did in 1844,[*] and then asserted their *fibrinous* composition, whence the term "fibrinous" cast still in use. From this time forward numerous communications appeared, and by 1845 the presence of casts in the urine of Bright's disease was an acknowledged fact.[†] In this year Scherer published Heller's theory, according to which casts were the result of congestion and exudation of liquor sanguinis.

To pass to more modern observers, Traube also held that casts are composed of fibrin, which is caused to exude by a high degree of intravascular pressure.

At the present day Dr. George Johnson[‡] says: "The basis of all renal tube-casts is the fibrin within the uriniferous tubules; but these casts assume various appearances, according to the nature of the products which they contain and the condition of the tubes in which they have been moulded."

Dr. Beale[§] says: "The transparent material probably consists of a peculiar modification of an albuminous matter possessing somewhat the same characters as the walls of some epithelial cells, the elastic laminæ of the cornea, etc., but not condensed like these structures." This, according to him, is the basis substance, which, becoming solid, entangles epithelial cells, nuclei, granular matter, blood-corpuscles, or whatever may be in the tube at the time the effusion occurs; or which solidifies in the shape of a hyaline cast, if the tube be empty, or the epithelium so firmly adherent that the cast slips out without detaching it.

Dr. Dickinson[||] believes that, as a rule, fibrin forms the basis

---

[*] Henle, in Henle und Pfeufer's Archiv, Band i, Heft i, s. 60, 1844. Henle, in Zeitschrift für rationelle Medizin, Bd. i, s. 68.

[†] See the work of Burkhart alluded to for a full historical sketch, more particularly of German authors.

[‡] Johnson, Lectures on Bright's Disease, New York, 1874.

[§] Beale, Kidney Diseases and Urinary Deposits, Philadelphia, 1869, p. 339.

[||] Dickinson, W. Howship, On the Pathology and Treatment of Albuminuria, 2d edition, London, 1877, p. 19.

of all casts, but says "it sometimes happens that cylinders are formed in the urine, which appear to consist entirely of compacted epithelial cells, or of epithelial cells held together by fibrin so small in amount as to be barely perceptible."

Dr. Grainger Stewart* says tube-casts are composed of co-agulated fibrin with altered epithelium and not unfrequently blood-corpuscles. Weissgerber and Perls† are among the most recent who adhere to the view that casts are formed of exuded fibrin, while Robin, as early as 1855, protested against the use of the term "fibrinous" cast, as being erroneous.

Axel Key,‡ having shown that the chemical characters of casts were incompatible with a fibrinous composition, early claimed that at least the so-called "hyaline gelatinous" and "hyaline waxy" casts are composed of desquamated epithelium which has fused and degenerated into a hyaline mass, and that dark granular casts are also formed by an agglomeration of degenerated epithelial cells. O. Bayer§ asserted that all varieties of casts originate in this melting together and transformation of cells. Schachowa,‖ in some experiments upon the artificial production of nephritis in dogs, by means of cantharides, obtained results which sustained the same conclusions. On the other hand, Œdmansson¶ claimed that all casts originate as a secretion from the epithelial cells lining the tubules, and Œrtels** defends the same view.

Langhans, in a recent article,†† says that while many casts

---

* Stewart, Grainger, Bright's Disease, 2d edition, New York, 1871, p. 80.

† Weissgerber und Perls, Archiv für Experiment. Pathologie, Bd. vi, 1876.

‡ Axel Key, Om. des k. Tubularafgjutningarnas olika former och bildning vid sjuckdomari Njurarne; Med. Archiv, Stockholm, 1863, I, 233–277, 2 pl.

§ O. Bayer, Ueber den Ursprung der sogenannt. Fibrin-cylinder des Urins, Archiv für Heilkunde, 1868, s. 136.

‖ Schachowa, Untersuchungen über die Niere. Dissert., Bern, 1876.

¶ Œdmansson, quoted by Bartels, op. citat., p. 84.

** Œrtels, Experimentelle Untersuchungen über Diphtherie, Deutsches Archiv für Klin. Med., Bd. viii, s. 292.

†† Langhans, Theodore, Ueber die Veränderungen der Glomeruli bei der Nephritis nebst einigen Bemerkungen über die Entstehung der Fibrincylinder, Virchow's Archiv, Bd. 76, erstes Heft, 1879, 85.

arise from a metamorphosis of cells contained in the uriniferous tubules, he is far from claiming that all casts originate in this manner. It is especially the waxy casts that are thus formed, while the pale homogeneous cylinders are a kind of secretion from the epithelium. According to Langhans, the cells which undergo conversion into waxy casts may be either the desquamated epithelium of the tubules, lymph-corpuscles, or red blood-disks. In either event, first melting into each other, they break up into a finely granular mass, which subsequently undergoes a glasslike transformation. This begins at the periphery, and extends thence towards the centre. A yellowish tinge and more highly refracting character result when the cells thus converted are red blood-disks.

Rovida,* whose observations on the nature of casts have been alluded to, makes but three varieties of casts,—the colorless, the yellow hyaline, and epithelial, of which he considers the colorless and yellow hyaline a secretion from epithelial cells.

Bartels† admits three modes of origin for casts. First, that dark granular casts originate directly by an agglomeration of degenerated epithelial cells, by which agglomeration I infer also he forms epithelial casts, although he makes no direct statement to that effect. He insists, however, that in every case in which he has examined microscopically thin sections of diseased kidneys whose tubules were blocked by the dark granular casts, the tubules invariably exhibited an epithelial lining. He reconciles this fact with his view by admitting that the theory of Key and Bayer, that the epithelium thus shed is rapidly reproduced, may be correct. Secondly, he believes that the yellow (waxy) casts arise as a species of secretion from the epithelium. Third, Bartels holds that certain "homogeneous, transparent, lightly streaked, or faintly shaded varieties, or the forms which are so delicately stippled with the finest granules or minute oil-drops,—in a word, the casts which are

* Rovida, Ueber das Wesen der Harncylinder, Moleschott's Untersuchungen zur Naturlehre des menschen und der Thiere, Bd. xi, s. 1.

† Bartels, op. citat, 84, 85, 86.

most rightly called hyaline, are formed by a coagulation of the albumen and its derivatives secreted with the urine." Their occurrence he considers entirely dependent upon the albumen mixed with the urine.

Rindfleisch* thinks the matter is not yet ripe for decision. He says that he has long favored the view that the epithelial cells of the straight tubes furnish a solidifiable colloid material in their protoplasm, which they pour out into the lumen of the tubules, and he gives a drawing in his work on *Pathological Histology* which he believes lends support to this view.

At the same time he says he is forced to admit, with Klebs, that this may be a post-mortem product, or the result of a method of preparation; also, that casts may be the result of a liquefaction and fusion of red blood-corpuscles; and finally, that for the present the matter must rest with the statement that a fluid albuminous substance is poured out into the lumen of the tubule, which coagulates, and that thereby the fibrin-cylinder is formed. In the next paragraph, however, under "amyloid infiltration," he says that in advanced stages of this condition, when alone the epithelium becomes involved, casts may be formed by the fusion of the infiltrated cells.

Charcot† is not at all explicit, but denies that they are fibrinous, and considers that some (certain granular casts) are made up of broken-down epithelial cells, others (hyaline and some granular) of an albuminous substance, while epithelial casts are agglomerations of epithelial cells more or less altered.

The Part of the Uriniferous Tubule in which Casts are Formed.

Casts may be formed in all parts of the tubules of the kidney, but it is commonly asserted that most of those found in the urine come from the looped tubes of Henle and the collecting-tubes, because from these escape is easier than from

---

* Rindfleisch, Pathological Histology, New Syd. Soc. Transl., vol, ii, 1873, pp. 144. Fifth German edition, 1878, 442, 443.

† Charcot, On Bright's Disease, translated by Millard, New York, 1878, p. 29 *et seq.*

the convoluted tubules; not, however, easier than from the *intermediary* or *intercalary* portion, in which each ascending limb terminates before it empties into a collecting-tube. This part of the tubule is identical in structure with the convoluted portion. Rindfleisch says that casts are formed only exceptionally in the convoluted tubules, and that when found *in situ* in the cortex of the kidney, it is in this intermediary piece of Schweigger-Seidel, and not in the convoluted portion, that they usually lie. While there are difficulties in the descent of a cast from the cortex of the kidney, and while such descent is doubtless more rare, I do not consider it impossible. The cast is smooth, flexible, and may contract its dimensions decidedly. Under these circumstances the *vis a tergo* of the newly secreted urine, which must be considerable, may be sufficient to force it downwards along the loop of Henle and through the intermediary piece into the collecting-tube. It is scarcely necessary to say that a blocking up of the convoluted tubules by casts is a very much more serious matter than of other portions, and that if it were of frequent occurrence a larger proportion of cases of Bright's disease would be immediately fatal.

It is not unlikely that those cases of chronic parenchymatous nephritis, with large albuminuria and very few casts, such as that of the young girl which will be further alluded to on page 76, are cases in which the convoluted tubules are exceptionally involved; for in that instance the cortex was crowded with casts, while the urine contained but few, and the fatal uræmia which terminated the case may have succeeded this involvement.

The Significance of Casts found in Urine.

It is said that hyaline casts are found in urine from healthy kidneys,* and that epithelial casts are found as a result of the administration of diuretics;† but while I have found, in a very

---

* See Charcot on Bright's Disease, New York, 1878. Charcot says this was first pointed out by M. Robin in 1855, and confirmed by Axel Key, Rosenstein, and many others.   † Ibid.

few instances, casts in urine in which there was at the same time no albumen, I have never found true casts in urine from what I considered normal kidneys. In cases of irritation of the bladder, ureters, and pelvis of the kidney, there occur what, following Dr. Beale, I call *mucous casts*, very long, delicate, hyaline, faintly striated, often branching casts. They are evidently formed in the largest collecting and excretory tubes, and are probably mucus. They are due to irritation, propagated from irritated mucous surfaces. Sometimes their resemblance to casts is even closer in consequence of precipitation upon them of granular urates or amorphous phosphate of lime. But this granular deposit, when present, incrusts everything which may be in the urine, epithelial cells, foreign matter, etc., besides forming a copious precipitate of its own.

It is not very unusual to find cases in which there have been albuminuria and casts, where, as improvement progresses, the albuminuria disappears, and casts are found in the urine for some time longer. I have observed this in several cases, and I suppose, if watched for, more frequent instances would be found. My friend, Dr. Milner, of Chester, Pennsylvania, having had his attention called to the subject by a publica-tion of my own, in which it was stated that during recovery from acute Bright's disease casts might be found in the urine some time after albumen had disappeared, examined carefully the urine of a child recovering from scarlatinal nephritis, and found casts for two weeks after albumen had disappeared. Such cases require no further comment. Bartels correctly says: "The formation and the excretion of a cast need be by no means contemporaneous events."

An adult woman, on admission to the Pennsylvania Hospital, exhibited a slight puffiness and pallor of countenance, which suggested an examination of her urine. No albumen was found by the physicians at the hospital, but a considerable number of slightly granular and hyaline casts. Dr. James H. Hutchinson sent me a specimen of the urine, and by no test could I discover the presence of albumen, but found delicate granular and hyaline casts. *One week later* a searching examination of the urine of the same patient revealed *neither casts nor albumen*. This last observation settled the question. The case had been one of Bright's disease which was convalescent. Finally, A. Henderson reports for Osler, in the *Canada Medical and Surgical Journal*, Montreal, 1879, vol. vii, a case of persistence of casts and blood-corpuscles in urine after albumen had disappeared. Now if we add cases of contracted kidney, in which occasionally for a time either casts or albumen may be wanting, this category is complete.

On the other hand, occasional instances of true renal albuminuria, and large albuminuria without heart disease, are sometimes met in which casts are absent. But even here, if the opportunity for repeated examinations is present, there comes a time when a few casts will be found, while later again an examination may fail to discover them. Such a case of supposed amyloid disease of the kidney I have now under observation, fully noted. I have notes of another case of copious albuminuria, in which three careful examinations failed to discover casts, and the patient subsequently died; but no post-mortem examination was made. My own relation to the case was confined to the urine examination. It may have been one of heart disease, but I believe it was one of parenchymatous nephritis. . Recently I was asked to see a young girl of 14, the daughter of a physician, who had copious albuminuria for nearly a year before I saw her. During this time her urine was examined for casts several times, but none were found. At my examination, although there was large albuminuria, *only a very few casts were found*. She died in a few days in uræmic convulsions, and the autopsy revealed two typical large white kidneys, weighing each about eight ounces. Upon microscopic examination of sections, the tubules of the cortex were found to contain numerous waxy casts. Finally, I am confident that there are many instances of albuminuria where the absence of casts is reported in which the examination is at fault, being carelessly performed.

From these remarks it may be inferred that I attach no small significance to the presence of casts in the urine. Although

I have never seen casts in the urine from healthy kidneys, I will not deny the possibility of its occasional occurrence. But when we remember that urine is seldom investigated with the microscope, unless we are searching for the cause of an evident or suspected ill health, we may lay it down as a rule, to which exceptions are so rare that they may be ignored, that when casts of the uriniferous tubes are found in a specimen of urine the kidneys whence they come are not in a normal state. That state may be temporary, but if so the casts pass away with it, while, if it is permanent, the casts continue during its existence. The doubtful cases are, of course, those unaccompanied by albuminuria. I presume no one will claim that the kidneys which shed both casts and albuminuric urine are healthy. Cases of the former category require, of course, to be studied, and the issue will always clear itself. I cannot comprehend the position taken by Charcot in his recent work, where, in addition to saying that " in a general way, the clinical importance of urinary casts has been very much exaggerated," he also says of epithelial casts, " it may be said that, from a clinical point of view, they do not possess much importance ;" and of hyaline casts, " even when renal disease exists, the hyaline casts are of no interest, except from their long persistence."

When we come to ask the question, whether by the kind of casts present we can diagnose the nature of the renal malady, a less positive reply can be given ; but, nevertheless, some of the varieties give us considerable assistance in making a diagnosis. The following general statements may be made :

1. Hyaline casts are found in all forms of Bright's disease, as well as in temporary congestions of the kidney, active or passive.

2. Epithelial casts are found in acute, subacute, and chronic parenchymatous nephritis. In the latter two forms the cells are generally degenerated and fragmentary.

3. Blood-casts are found in acute parenchymatous nephritis, and where hæmorrhages have occurred in the kidneys.

4. Pale granular casts are found in interstitial nephritis

(contracted kidney) and chronic parenchymatous nephritis (large white kidney).

5. Dark granular casts are found in parenchymatous nephritis, acute and chronic, and rarely in interstitial nephritis.

6. Waxy casts are found only in chronic Bright's disease, and attend either of the three principal forms.

7. Oil-casts are found in subacute and chronic forms of Bright's disease, and may attend any of the three principal forms, but are most numerous in chronic parenchymatous nephritis.

8. Free fatty cells and free oil-drops are found in chronic parenchymatous nephritis.

9. The form of fatty cell known as the compound granular cell is found in acute and chronic parenchymatous nephritis.

SECTION IV.

CLASSIFICATION OF BRIGHT'S DISEASE.

WHATEVER may be the objections to retaining the term Bright's disease or Bright's diseases in medical nomenclature, there is every reason to believe it will remain in use for many years, if not as long as medical science exists. This being the case, I see no reason for excluding from the category any one or more of those affections of the kidney attended by albuminuria, dropsy, and tube-casts, either simultaneously or in close relation to each other. For, accurately described and delineated by Bright as were some of these forms of disease which he studied, it is evident that he had no correct notion of the intimate nature of the processes which constitute them; while it is almost certain that among those cases of dropsy and albuminuria which came under his notice were examples of forms of disease which certain authors would eliminate from the category of Bright's disease. Thus it is scarcely likely he could have failed to have had under his observation in a large hospital experience cases of the so-called lardaceous or amyloid disease of the kidney, though he failed to recognize it as a separate form, and probably included it under one or another of the three described by him.

Again, great pains have been taken to draw a sharp distinction between Bright's disease and the structural changes which take place in the kidney as the result of passive congestion, such as obtains in valvular disease of the heart, although, as will be shown later, the histological results are the same as those which in a much more advanced stage constitute the granular contracted kidney of Bright. I would, however, retain even this condition under the category of Bright's disease, notwithstanding the fact that the alterations seldom reach even the slighter degrees of contraction essential to the morbid anatomy of the contracted organ. Traube and Kelsch are, so far as I know, alone in desiring to discard the term Bright's disease altogether.

Bright's* first paper appeared in 1827, and of the earlier writers after him, Rayer was the first to assign an inflammatory character to the changes in the kidney described by the great English physician. Rayer also said of the six forms which he himself described that they are "probably successive." Rokitansky (1842) was the first to include lardaceous disease in the category of Bright's disease.

Reinhardt (1850) and Frerichs (1851) after careful histological investigations also concluded that the conditions described by Bright were the result of inflammation, and sought to establish the proposition of the "unity" of Bright's disease, that is, that the different forms represent three successive stages of this process: 1st, hyperæmia; 2d, exudation with fatty degeneration of epithelium : and, 3d, new growth of connective tissue with resulting atrophy of the kidney.

These observers laid especial stress upon the casts found in the urine during life and in the uriniferous tubules after death. Supposing their fibrinous nature established, they considered the inflammation a "croupous" one, and applied the term "diffuse croupous nephritis" to the entire process on account of its uniform distribution over the organ. Finally they included under Bright's disease all affections of the kidney attended by casts; among them, therefore, the results of passive congestion, or, as they are now called, cyanotic induration and lardaceous disease. Traube (1856) first insisted that the former was a distinct process and should not, therefore, be included under Bright's disease. He also showed how lardaceous disease could be distinguished from other renal diseases; and although Bamberger earnestly strove to replace the former in the category, the views of Traube have since generally prevailed in Germany.

---

* Bright, Richard, Report on Medical Cases, London, 1827.

Virchow (1852), in his famous paper on parenchymatous inflammation,* makes, first, a milder degree of inflammation, which he terms *catarrhal* nephritis, in which there is first, a hyperplasia of epithelial cells, and later an alteration and shedding of the same; second, a *croupous* inflammation, wherein a fibrinous exudation is added to the catarrh; and, third, a *parenchymatous* inflammation, in which, too, the cells are altered, swollen, cloudy, granular, and subsequently break up. To what he considers the very rare combination of these three condition,—catarrhal, croupous, and parenchymatous nephritis,—he suggests that the term Bright's disease, if used at all, be restricted. In his *Cellular Pathology* (1858), however, Virchow includes under the term Bright's disease three conditions, one of which originates in the tubules (parenchymatous nephritis), a second in the intertubular tissue (interstitial nephritis), and a third in the vessels (amyloid degeneration). Thus Virchow was the first to call attention to the intertubular connective tissue as the special seat of a change. To this he had, in a more general way, alluded in his earlier paper. Arnold Beer† and Traube,‡ on the other hand, located the entire primary process in this tissue, making the changes in the epithelium altogether secondary, and do not acknowledge such a process as parenchymatous nephritis.

Klebs (1870), in his *Handbook of Pathological Anatomy*, excludes cyanotic induration and lardaceous disease from *morbus Brightii*.

Rosenstein,§ in his work on *Diseases of the Kidneys*, treats of congestive hyperæmia, catarrhal nephritis, diffuse nephritis, and amyloid degeneration, and says these are what have commonly been included under the term Bright's disease. But he himself uses the term as a synonym for his "diffuse

---

* Ueber Parenchymatöse Entzündung, Berlin, 1852.

† Beer, Die Bindesubstanz der Menschlichen Niere, Berlin, 1859.

‡ Zur Pathologie der Nierenkrankheiten, Gesammelte Beiträge zur Pathologie und Physiologie, 2 Bd., 2 Abtheilung, Berlin, 1860.

§ Die Pathologie und Therapie der Nierenkrankkeiten. Zweite verbesserte Auflage, Berlin, 1870, ss. 51, 100.

nephritis." He adopts the older view of the "unity" of the disease, and by his diffuse nephritis he means what most writers of the present day call a parenchymatous nephritis, that is, a nephritis in which the tubules and their epithelium are the chief centre of change. Of this, however, he makes three stages, after the scheme of Reinhardt and Frerichs, a stage of hyperæmia, one of exudation, and one of atrophy, the result of which is the granular contracted kidney. Thus atrophy may result from the parenchymatous process alone, by a fatty degeneration and breaking down of the cells; but is, for the most part, accompanied by changes in the interstitial tissue, which consists sometimes in a fibrillar and sometimes in a cellular hyperplasia, whose contraction co-operates to produce the atrophy of the organ. Rosenstein's catarrhal nephritis corresponds with Virchow's, that is, it is a slight degree of parenchymatous inflammation. Of interstitial nephritis as a special form he makes no mention.

The English authors, and notably among them Dr. George Johnson, were most instrumental in breaking up the notion of the "unity" of Bright's disease for a long time so stubbornly held by the German and French pathologists, and still maintained by Rosenstein and some others. Among those who have emancipated themselves most thoroughly from this doctrine are Bartels and Charcot. The former, in his recent article in Ziemssen's *Cyclopædia of Medicine* (1877), says that among the diseases originally grouped under the title Bright's disease, his clinical experience and pathologico-anatomical investigations have led him to include hyperæmia, acute and chronic, ischæmia, parenchymatous inflammation, acute and chronic, interstitial inflammation or connective tissue induration, and amyloid degeneration. He also announces his firm allegiance to the English view of the nature of the disease. Charcot says:* "In my opinion, therefore, which is only, so to speak, the reflex of the English theory, Bright's disease is a class comprising several distinct species, not only from an anatomico-

---

* Charcot on Bright's Disease, translated by Millard, New York, 1878, p. 36.

pathological point of view, but as regards etiology and symptomatology." These distinct species are parenchymatous nephritis, interstitial nephritis, and the amyloid kidney. Of modern French writers, Cornil and Ranvier, Lancereaux, Lecorché, Labadie-Lagrave and others, hold the same views.

Of the English writers above referred to, Dr. Johnson first promulgated his views in his larger work on *Diseases of the Kidney*, published in 1852, but his most recent views are found in his smaller treatise, entitled *Lectures on Bright's Disease*, published in 1873. He describes an acute and chronic form. The former is represented by acute nephritis, the latter by the red granular kidney, the large white kidney, and the lardaceous or waxy kidney. Roberts, Dickinson, and Grainger Stewart, while differing somewhat as to the exact pathology of the different forms, adopt practically the same arrangement. Beale agrees entirely with Johnson. Grainger Stewart and Dickinson are the most systematic, and attempt a sharper differentiation of the different forms of parenchymatous nephritis. The former writer makes:

1. The *inflammatory* form, of which there are three stages, that of inflammation, that of fatty transformation, and that of atrophy.

2. The *waxy* or *amyloid* form, of which there are also three stages, that of degeneration of vessels, that of secondary changes in the tubes, and that of atrophy.

3. The *cirrhotic, contracting,* or *gouty* form.

Dickinson makes:

1. Acute tubal and diffuse nephritis, involving primarily the tubules, but extending, sooner or later, to the interstitial tissue, terminating in fatty degeneration and subsequently contraction, if early death or recovery does not take place.

2. Granular degeneration.

3. Lardaceous disease.

In the following pages I shall retain the term *Bright's Disease*, which I subdivide into *acute* and *chronic*.

I. Acute Bright's disease is represented by a single form, *acute* parenchymatous nephritis.

II. Chronic Bright's disease includes the following :

1. Chronic parenchymatous nephritis.

2. Lardaceous disease, a chronic process.

3. Interstitial nephritis which is only chronic.

As *acute hyperæmia* is represented by the first stage of parenchymatous nephritis, and is not characterized by a clinical history essentially different, I will not devote to it a separate consideration.

In like manner *cyanotic induration*, the pathological product of passive congestion, is the result of a process similar to that of interstitial nephritis, with the first stage of which it might, with some propriety, be considered to correspond. But in consequence of its relation to valvular disease of the heart, which is its immediate and almost exclusive cause, while there are also some symptoms peculiar to it, I will devote a separate section to its consideration.

Finally, to secure completeness, I shall devote a section to *suppurative interstitial nephritis*, or pyelonephritis, without desiring to be considered as including it under Bright's disease.

SECTION V.

ACUTE PARENCHYMATOUS NEPHRITIS.

Synonyms —Acute nephritis, acute desquamative nephritis, acute tubal nephritis, acute Bright's disease, acute catarrhal nephritis, croupous nephritis, albuminous nephritis, haemorrhagic nephritis (Traube), acute albuminuria, acute renal dropsy.

ACUTE PARENCHYMATOUS NEPHRITIS IS AN ACUTE IN-FLAMMATION OF THE KIDNEY, IN WHICH THE EPITHELIUM OF THE TUBULES IS THE PRIMARY AND PRINCIPAL SEAT OF INFLAMMATION.

Etiology.

Most cases of acute parenchymatous nephritis are caused by *scarlet fever*, and occur, therefore, in children. A certain number originate in exposure to cold, especially cold and moisture, while the body is warm and perspiring. The latter cause is particularly efficient if the person be fatigued or exhausted.

As to the method in which *scarlatina* causes the nephritis, while it cannot be said to be undisputed, it is most likely that the specific poison of the disease operates through the blood upon the epithelium of the kidney, which probably seeks to remove it. In evidence to support this view is the fact that it is by no means the cases in which the eruption is most profuse, and in which the skin is therefore most sensitive, that the renal complication occurs; also that there are specific agents which, when introduced into the blood, produce identical results. Such agents are cantharides, turpentine, and oil of mustard. The independence of cold, of certain cases of nephritis following *scarlatina*, is further attested by those instances, familiar to every practitioner of experience, in which the disease has succeeded upon *scarlet fever* during a convalescence in which the patient has been kept in bed, while the barefoot pauper may have run the streets with the eruption upon him, and throughout convalescence, with utter impunity. But while it cannot be claimed that cold, by causing the retrocession of blood from the skin to internal organs, can alone produce the nephritis of scarlatina, it cannot be denied that it may co-operate with the peculiar poison in producing the same result. And in those cases of acute nephritis which result from exposure to cold, the immediate cause is probably a congestion due to the introcession of the blood from the surface of the body; but here again, it is not unlikely that the supplemental function assumed by the kidney in consequence of the suppression of excretion by the skin, contributes secondarily. The same may be said of the acute nephritis which succeeds upon extensive burns of the surface of the body.

In like manner, other grave infectious diseases, as *diphtheria* and *small-pox*, cause acute parenchymatous nephritis, the former frequently. In both it presents itself, if at all, at the *acmé* of the disease. *Acute endocarditis* and *acute articular rheumatism* are also occasional causes; while *measles, erysipelas, pyæmia, jaundice*, and *diabetes* have been known to cause it. In a case of chronic Bright's disease under my care a sharp acute attack was evidently induced by an attack of *erysipelas;* and a pair of kidneys, removed post-mortem from a case of *diabetes* and sent to me for examination, presented all the signs of an acutely congested kidney. The alterations which take place in the kidney after *cholera* are generally acknowledged to be those of inflammation. *Skin diseases*, as well as *extensive burns of the skin*, are acknowledged causes; the former rarely, but the latter almost always if the burns be sufficiently extensive.

Pregnancy is the cause of a good many cases of acute parenchymatous nephritis, indeed so important a cause of it that some writers, as Bartels, devote a separate section to the acute parenchymatous nephritis of pregnancy.

Most cases of acute nephritis due to other causes than scarlatina, cold, and pregnancy, are mild in degree; and even in cases due to pregnancy, if the patient is once through with her confinement, recovery is usually rapid.

In looking for the evidence of nephritis in acute infectious diseases it must not be forgotten that *intense febrile move-ment* may cause albuminuria, independently of any structural change in the kidney due to the zymotic agent. When thus caused the albuminuria is always small.

When acute nephritis supervenes upon scarlet fever it does not usually make its appearance until the end of the second week. Indeed it is often during supposed convalescence that it most unexpectedly occurs. Bartels, in a single case, detected it on the tenth day, and never later than the thirty-first. He found the twentieth day to be the mean limit, while the great-est number of cases also presented themselves on this day.

Dr. Tripe* concludes from his observations that dropsy may come on at any period of scarlatina, even the earliest, but that it most frequently appears on the fourteenth day. Dr. West says it is most frequently in the second week that it occurs, and that if delayed later the symptoms are generally milder. Dr. Dickinson† found the majority of cases occurring in the third week, but says: "Speaking generally, it may be said that after the end of the first month the danger is small, but until after the lapse of the second the patient cannot be looked upon as safe." These are safe limits. I cannot but think that the earliest cases of Dr. Tripe—those which oc-curred say within the first week—may have been cases of albu-minuria resulting from the intensity of the febrile movement, which of itself may cause albuminuria though not nephritis.

Certain specific *poisons* of vegetable and mineral origin are capable of producing acute nephritis. *Alcohol* is one of these. Although by no means so frequent a cause of Bright's disease as formerly supposed, undoubtedly a few cases are directly traceable to it, in one instance, quoted by Dr. Dickinson from Dr. Goodfellow,‡ apparently to the inhalation of its vapor. Among the best known of these substances are *cantharides, turpentine, oil of mustard,* and *phosphorus;* in a less degree the mineral acids, *arsenic, nitrate of silver, lead,* and *mercury.*

---

* Tripe on Scarlatinal Dropsy, Med.-Chir. Rev., 1854–55.

† Dickinson, op. citat, p. 90.

‡ Dr. Goodfellow on Diseases of the Kidney, p. 177.

I think *wormseed oil* may be placed in the same category, for in a case which came under my observation, an adult male died twenty-four hours after taking an ounce of this oil, death being preceded by coma and convulsions, while the urine was found to contain half its bulk of albumen.

Age and Sex.—As may be inferred from the etiology, acute nephritis is a disease of early age, although when caused by cold or any one of the causes named except scarlatina, it is as more likely to affect adults as these latter are more frequently subjected to such causes.

In one instance only have I known acute nephritis to originate in a person over thirty years of age. The patient was a woman, and the cause exposure. The possibility of its later occurrence cannot be denied. Dr. Dickinson says it is rare after forty, almost unknown after fifty. The oldest child I ever knew to have scarlatinal nephritis was a girl of fourteen, who died.

More males are attacked than females in adult life, evidently because they are more frequently exposed to the causes. But even in childhood there is a slight preponderance of cases in boys affected, which can hardly be thus accounted for.

Morbid Anatomy.

This varies somewhat with the stage of the disease as well as its severity. In the first place, as ordinarily caused, it is symmetrical, both organs being alike involved. The alterations may be so trifling as not to be recognizable by the naked eye. But the kidneys are generally sooner or later enlarged, in the latter stages always, sometimes to more than twice their normal volume, and they may weigh from eight to twelve ounces, those of children reaching the former, and those of adults the latter.

The capsule is distended by the enlargement, and, therefore, gapes when incised.* It may be otherwise unaltered or slightly

---

* Dickinson reports a case in which the capsule of both kidneys burst from the extent and suddenness of the tumefaction. Dickinson, Albuminuria, 2d edition, London, 1877, p. 97.

injected, and strips off easily, without dragging any of the parenchyma with it. Bereft of its capsule, the kidney itself is softer
than in health, inelastic, doughy. Its surface is smooth and exhibits a peculiar mottled appearance, which is due to the fact
that the little circlets of veins which form the boundary of the
lobules are distinctly injected, while the area included by each
circlet is paler than in health, and in the more advanced stages
even yellowish-white in color. This "irregular mixture of
congestion and anæmia," as Dr. Johnson calls it, is further contributed by the injection of other veins, indistinct in health.
The stellate veins, which are also more or less injected, are not
so conspicuous as in chronic parenchymatous nephritis. Spots
of hæmorrhagic extravasation are also found scattered over the
surface.

On section it is evident that the enlargement is due to
change in the cortex and the interpyramidal convoluted portion.
The cut surface is smeared over with a dark-red or chocolate-
hued blood, but on scraping or washing it away, the vessels are
found similarly injected to those of the surface, and between
them the same paleness or yellowish-white hue. The Malpighian bodies appear as distinct dark-red dots, and any linear vessels are distinctly filled, while punctiform hæmorrhages
may again be present as on the surface of the organ.

The *pyramids* are usually not essentially altered in appearance. They may be congested, and this may increase the vivid
contrast already existing between their dark-red color and the
pale cortex. In cases of extreme swelling, the pyramids may be
compressed in their central portions by the interpyramidal convoluted structure, which shares the enlargement of the cortex
of which it is an extension, producing the very characteristic
appearance first compared by Rayer to a wheat-sheaf.

Minute Changes.—These are confined almost solely to the
cortex. They begin at least in the epithelium, and vary a
good deal with the stage of the disease. The earliest condition of the cells is undoubtedly that of cloudy swelling,
the result of increased nutritive activity. In this state the
cells are swollen, slightly more cloudy than usual, in conse-

quence of a deposition of albuminous granules in their interior, which may accumulate to such extent as even to obscure the nucleus. Although kidneys removed after death from cases of acute parenchymatous nephritis have as a rule advanced far beyond this stage, yet it is often possible to find points less advanced at which cloudy swelling exists alongside of more advanced stages, while alongside of these again may be tubes in which the epithelium is healthy. As a result of the cloudy swelling, the cells are larger, and the tubes are therefore broader than they are in health, but a stage later the widening is more marked and the tubes are filled with more highly as well as moderately granular cells, to which are added free granular matter and often red blood-corpuscles. Under a low power the tubules appear as black, more or less opaque lines. The granules result from the breaking down of the cells, and the blood-disks are, of course, derived from the capillary vessels.

A closer examination of the *cells* at this stage, as obtained by scraping or examined *in situ*, shows them to be granular in various degrees. In some the nucleus is still visible, in others demonstrable by the aid of staining fluids, but otherwise invisible, and in others still, entirely obscured. Occasionally a few fat-drops may be present. In other situations the cells are so closely packed in the tubules that they cannot be differentiated, are apparently fused in one continuous, dark granular mass. This is the result of a hyperplasia of cells, and it is to these tubules thus distended with granular cells and their debris, dark by transmitted light, but white by reflected, that the pale or white color seen between the injected bloodvessels owes its origin.

Minute extravasations of blood visible to the naked eye have been referred to. These are found to occupy the tubules, and come either by diapedesis or rupture, from the capillaries of the Malpighian glomeruli, which, in the fresh condition, are also distended with blood-corpuscles. In a still more advanced stage, however, the Malpighian bodies may be paler, in consequence of the compression exerted by the still more swollen tubules, or by the proliferated cells peculiar to the capsule and its glomerule (glomerulo-nephritis).

Casts of the uriniferous tubules are also often found *in situ.* These are either blood-casts or small hyaline casts. These hyaline casts are fibrin, and when after contraction they are expelled from the tubule they may carry the epithelium of the tubule with them, thus producing a so-called epithelial cast.

Oertel claims that in renal diseases following diphtheria he has found "great numbers of *micrococci* and exuberant proliferations of the same," both in the renal tubes and Malpighian bodies. Heller* alleges he has repeatedly found the bloodvessels and their branches in acutely inflamed and swollen kidneys from cases of pyæmia, greatly dilated and plugged with masses, which under low powers presented a peculiar grayish-yellow appearance, and with higher powers were found to consist of extremely minute, highly refracting granular particles, placed at equal distances from one another. These particles he considered *spherical bacteria,* and the resulting masses *bacteria-emboli.*

The epithelium lining the tubules of the straight tubes is unchanged, but the tubes themselves often contain the same cellular and granular material contained in the convoluted tubes, which has descended from the latter.

The Interstitial (intertubular) Tissue of the Kidney in Acute Parenchymatous Nephritis.—The changes above described, it is seen, belong to the tubules and bloodvessels. In most cases of acute parenchymatous nephritis, there is no interstitial change, no formation or deposit of new material between the tubes. This assertion may be reaffirmed, but requires some further comment. It is certainly not present in the earliest stages; but sooner or later such changes do present themselves, and it is a matter of duration of the disease as to whether they appear or not. If the disease continues without permanent amendment for three months or longer, they will certainly have made their appearance. But here the border between acute and chronic parenchymatous nephritis has been passed. It is impossible to say, however,

---

* See Bartels, in Ziemssen's Cyclopædia of Medicine, vol. xv, 1877, p 272.

precisely when they make their appearance, and for some time before such changes are recognizable by the naked eye they are discernible by the microscope. They consist in a hyper-nucleated overgrowth resulting from the proliferation of the connective tissue corpuscles always present, or from the fixation of wandered-out colorless blood-cells, and are doubtless due to the long-continued hyperæmia. It is possible, too, that there may be cases of such extreme severity that a diffuse nephritis, involving both the tubes and intertubular elements, may exist from the very beginning. Dickinson figures a section of a kidney showing such involvement, from a case of diffuse nephritis of ten weeks' standing in a boy of seven.*

Changes in the Pelvis of the Kidney. The mucous membrane of the kidney may be injected, but is otherwise unchanged.

Alterations in the Glomerule and its Capsule.

I have decided to describe here the pathological alterations of the Malpighian body, known as glomerulo-nephritis, because the changes are most frequently found associated with acute nephritis, although not exclusively.

The earlier writers on Bright's disease very generally described changes in the elements of the Malpighian body in both the acute and chronic forms. Thus Foerster† speaks of proliferation of the capillary nuclei, thickening of their walls, and desquamation of the enlarged epithelial cells lining the capsule. Virchow‡ described a cloudiness and increased number of nuclei in the capillaries of the Malpighian tuft in chronic nephritis, together with a thickening of their walls and greater width of the capillary coils. After these observers no allusion seems to have been made to such changes until Klebs published in his *Handbook of Pathological Anatomy*, in 1870, an account of some further changes in this body in acute ne-

---

* Dickinson, op. cit., p. 28.
† Foerster, Pathol. Anat., Bd. ii, s. 512.
‡ Virchow, Gesammelte Abhandl., s. 485.

phritis after scarlet fever. These he included under the title *glomerulo-nephritis*, and ascribes them to a proliferation of the nuclei in the *interstitial tissue* of the glomerule, the existence of which Axel Key had previously asserted. In this condition, according to Klebs, the Malpighian bodies appear to the naked eye as little white bloodless points. The urinary tubules are often not at all altered, though sometimes the convoluted tubules are slightly cloudy. The microscope discovers neither proliferation of epithelial cells (the so-called renal catarrh) nor interstitial alterations—nothing but the phenomena of congestion, if the Malpighian bodies are ignored. But the cavity of the Malpighian capsule is filled with small angular nuclei imbedded in a fine granular mass which almost completely covers the vessels. This, Klebs says, is not the endothelial lining of the capsule, because, on careful dissection by needles, this endothelium is found very slightly altered; ordinarily clearer and more firmly adherent than in health, occasionally fattily degenerated. He believes it is the compression of the vessels of the Malpighian body by this hyperplasia which sometimes causes the sudden, almost total suppression of urine and acute dropsy in cases of scarlet fever, followed by uræmia and death in from 12 to 24 hours.

Johnson[*] remarks that the nuclei in the walls of the capillaries of the Malpighian body are abnormally conspicuous. Birch-Hirschfeld[†] describes a proliferation of nuclei between the vascular loops and the epithelium; Cornil and Ranvier[‡] a swelling and granular condition of the capsular epithelium and nuclear proliferation and fatty degeneration of the capillary wall. Bartels[§] gives a drawing by Colberg of a glomerule from a case of chronic parenchymatous nephritis, in which the capillary nuclei are in a state of proliferation. And Litten[||] has

---

[*] Johnson, Lectures on Bright's Disease, New York, 1874, p. 30.

[†] Birch-Hirschfeld, Pathol. Anat., Leipzig, 1877, p. 1021.

[‡] Cornil and Ranvier, Precis d'Histologie Pathol., Paris, 1876, p. 1026. American Ed., 1880, p. 616.

[§] Bartels, in vol. xv, Ziemssen's Cyclopædia, New York, 1874, p. 373.

[||] Litten, Charité Annalen, t. iv, p. 30.

described marked proliferation and desquamation of the epithelium of the glomerulus itself, as well as of the capsular epithelium in a case of scarlatina.

But Langhans* has quite recently furnished the most complete and thorough account of the changes in the Malpighian body yet published. I have already made use of Langhans's paper in my account of the histology of the Malpighian body. It will be remembered that there are, according to him, three sets of cells contained in it, viz., the cells forming the lining of the capsule, the epithelial cells covering the glomerule or capillary coil, and the nuclei of the capillary vessels. Langhans was unable to convince himself of the existence of the stellate elements which Axel Key described as connective tissue cells, nor of the presence of any connective tissue within the capsule, except the adventitia of the vas afferens, which extended only as far as the branching of this vessel. But of the three sets of cells described by him all share in the pathological processes of the kidney.

First, as to the *epithelium of the capsule*, it exhibits in its slightest degrees of alteration, such as occurs in a congestion of the kidney, a simple overgrowth of the cells, so that instead of being flat they are more comparable to the epithelial lining of the tubules in their convoluted portion. The nucleus is larger, oval, and fills the cell almost completely. This degree of alteration occurs in any of the forms of Bright's disease. Increase in the number of cells, resulting in stratified layers, is a much more rare occurrence, but Langhans has seen it in *acute* nephritis where there were hæmorrhages within the capsule. Here the Malpighian body is elongated in a direction parallel with the axis of the medullary rays. This is due to the presence between the capsule and the glomerule of a crescentic mass of cells, of which the thickest portion is opposite, or nearly opposite, the point of entrance of the vas afferens. In the spaces between the layers of these cells, which are more

* Langhans, Theodore, Ueber die Veränderungen der Glomeruli bei Nephritis nebst einigen Bemerkungen über die Entstehung der Fibrin-cylinder, Virchow's Archiv, Bd. 76, 1879, s. 85.

or less epitheloid in shape, are found smaller and larger lymphoid cells. More frequent than this extreme degree is partial thickening of the capsular epithelium.

FIG. 30.

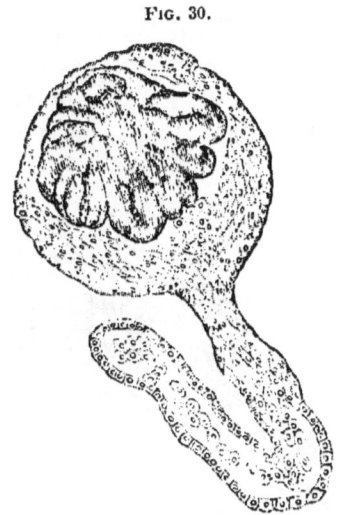

Proliferation and thickening of the capsular epithelium with compression of the glomerule. × 120.—After LANGHANS.

With regard to the effect of this proliferation upon the function of the kidney it is not unlikely that by exerting a pressure upon the capillaries of the glomerule it diminishes the quantity of urine secreted, especially as in Langhans's cases such a diminution occurred, and especially also as proliferation of the interstitial tissue in contracted kidney is attended by an opposite result; and alterations of the epithelium of the tubules are not necessarily attended by such results. The hæmorrhages which sometimes occur in the Malpighian body where these alterations exist, Langhans would locate, not in the capillaries of the glomerule itself, but in those of its capsule, since the blood is not found on the surface of the glomerule, but between the layers of the thickened capsular epithelium.

Next, as to alterations in the *glomerule-epithelium*, it is found that in inflamed kidneys the epithelium is not so readily removed in shreds, but rather in isolated cell elements; in other words, the cement substance uniting the cells seems to be dissolved, and the cells are probably also less closely adherent to the capillary loops. The cells themselves are very little altered, in some instances swollen; and when this is the case, the thicker nucleated portion of the cell is exclusively involved, and forms prominent button or clublike processes, which are attached to the convexity of the capillary loop, sometimes by a broad base, and sometimes by a delicate

Fig. 31.

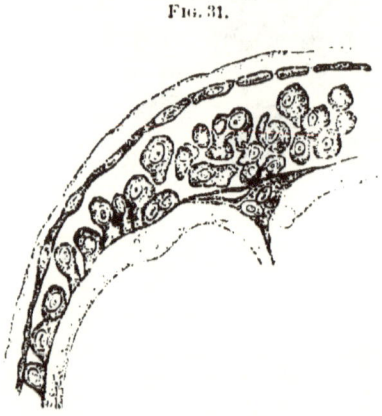

Desquamative glomerulo-nephritis. × 300. The surface of the capillaries, the details of which are omitted in the drawing, is covered with numerous cells.—After LANGHANS.

pedicle. This may separate completely, leaving the free nucleated portion as an independent cell between the capsule and the glomerule. Although the number of these is usually small, they may be so numerous as to enlarge and change the shape of the Malpighian body to an oval whose longer axis is parallel with the medullary rays and nearly .3 mm. ($\frac{1}{85}$ inch) in length. These cells are even sometimes found between the lobules of capillary vessels which make up the glomerule so as quite distinctly to separate them.

This condition, so far as the swollen state of the cells is concerned, is not infrequent. Langhans has never failed to see them in connection with the large white kidney. He also found it in a single case of acute nephritis along with marked hæmorrhages from the glomerule and conspicuous thickening of the capsular epithelium; also in a slight degree in granular atrophy of the kidney. On the other hand, in five cases of scarlatinal nephritis he failed to find any considerable swelling of the glomerule-epithelium. A single case of large white kidney, with hypertrophy of the left ventricle, attended during life by a very copious (4000–5000 cc., 133–166 f$\mathfrak{z}$) highly albuminous urine and numerous casts. Langhans thinks a good deal of light has been shed upon it by this condition of the glomeruli, in which alone the case differed histologically from other large white kidneys. To this he thinks may be referred the extraordinary increase in the quantity of urine, while the condition itself may be compared to a desquamative catarrh of the mucous membrane, which is in like manner attended by the exudation of a highly albuminous exudation. He does not, however, venture a decided opinion, since he lost the most favorable opportunity for investigating the condition of the capillaries. I cannot myself comprehend why such a proliferation should increase the quantity of urine, but would rather expect that it would operate similarly to the overgrowth of the capsular epithelium to compress the glomerule, and thus diminish the quantity of urine secreted.

Finally, as to changes in the *capillary nuclei*, according to Langhans these are by no means so rare as the silence of modern literature on the subject would lead one to suppose, although for their recognition more careful investigation is necessary than for the recognition of either of the two already considered. To this end complete isolation of the capillaries from each other and from the epithelium is essential, and is accomplished by means of dissection with and without artificial injection of the vessels. Even by low amplification a difference is observed between these Malpighian bodies in which the capillary nuclei are altered and those in which there is a proliferation of the epithelium. The glomerule is enlarged (.2 to .35 mm., $\frac{1}{125}$ to $\frac{1}{70}$ inch), and fills closely the entire cavity of the capsule, appearing, especially in the fresh condition, as a compact intensely clouded mass. But in hard-

7

ened preparations its peculiar subdivisions are easily recognizable. The epithelium of the glomerule as a rule exhibits simply swelling, especially the nuclear portion, as well as a marked extension of the single cells whose convexity has a radius considerably longer than normal, corresponding to the increased lumen of the capillary. The capillaries themselves are increased in diameter to .03 mm. ($\frac{1}{833}$ inch). After removal of the epithelium of the glomerule, the lumen of the capillaries is found more or less occupied by a cloudy, finely granular substance, which sometimes contains a few and sometimes a large number of minute fat-drops, takes up carmine, clears up and swells somewhat on the addition of acetic acid. In it are found numerous small round nuclei, .006 to .008 mm. ($\frac{1}{4168}$ to $\frac{1}{3125}$ inch) in diameter, exactly similar to the normal capillary nuclei, and quite different from the large oval epithelial nuclei. They are separated from each other by a distance equal to half their own diameter. That

FIG. 32.

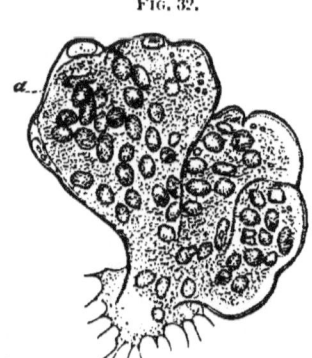

Capillary loops with proliferation of the nuclei, the epithelium being retained only at *a*.
× 300.—After LANGHANS.

these nuclei truly lie in the finely granular mass and are not adherent to the capillary wall, is seen at points of rupture, or still better on pressure, when nuclei and fine granular matter well-up without breaking up into single cells. These capillaries are still pervious to blood, but a higher pressure than usual is required to fill them, even then with but partial success. The extreme resistance results in a dilatation of the vas afferens outside of the Malpighian capsule and of the afferent arteries. Langhans found the diameter of the former vessel from a boy twelve years old who died of acute nephritis, .06 mm. ($\frac{1}{416}$ inch) in diameter, the normal in an adult being .014 to .02 mm. ($\frac{1}{1785}$ to $\frac{1}{1250}$ inch). From this resistance also result extravasations of blood into the capsule through the walls of the first branches of the vas afferens. Later, this finely granular mass appears to fuse more intimately with the membrana propria of the capillary wall and acquires the same optical properties as it. This process, although apparently independent of the nuclei of the capillary walls, Langhans calls " proliferation of the capillary nuclei." He also remarks that evidence is wanting as to whether the colorless corpuscles share in the process.

As to the circumstances under which these alterations occur, Langhans has never found them absent in the large white kidney, all the glomeruli being affected. He has also found them in acute nephritis, in one case along with proliferation of the capsular epithelium. Upon scarlatinal nephritis he had, however, made no observations. In granular atrophy the alteration is absent as a rule; it may be present, however, as a stage preliminary to contraction.

The influence of this nuclear proliferation of the capillaries upon the secretion of urine is difficult to determine in consequence of important alterations in the other elements of the kidney. It is very probable that it may cause a diminution in the quantity of urine secreted, and in this Bartels agrees with Langhans.

After carefully examining this subject I cannot but think that the proliferation of capillary nuclei is the same condition as that described by Klebs as a proliferation of the nuclei of the interstitial connective tissue of the glomerule, but which Langhans has been unable to discover. Klebs's drawing in his *Pathological Anatomy*, p. 646, is very coarse and difficult of interpretation. Both authors describe a nuclear proliferation imbedded in a granular matrix, but while Klebs puts it outside of the bloodvessels, Langhans puts it within. The question is by no means an easy one to settle, and I would not pretend to decide which is correct. I have carefully studied preparations in which this nuclear proliferation in the glomerule was marked, but by the optical means at my disposal could not decide upon their exact seat. Reasoning from the fact that they were accompanied by a general nuclear prolif-eration in the situation of the connective tissue elsewhere in the kidney, it would appear likely that they should occupy it here in the glomerule. But Langhans's work seems to have been carefully done, and the difficulty of artificial injection as well as the dilatation of the afferent vessels point to the cor-rectness of his views. Further researches are necessary to settle the question. In either situation the proliferation would necessarily have the effect of diminishing or suppressing the secretion of urine.

Symptoms and Course of Acute Parenchymatous Nephritis.

The mode of onset of acute nephritis is not uniform, but among the symptoms earliest noticed is slight swelling or puffiness in the face, below the eyes. This œdema rapidly extends to the upper extremities and trunk, and thence, if the disease does not abate, into the lower extremities and ab-dominal walls. In the male, the scrotum and prepuce are favorite seats of swelling. I have known the latter in a little child to be so great as to obstruct the passage of urine and re-quire catheterization, and that of the scrotum to result in sloughing. The great serous sacs are the last to fill with fluid in acute nephritis, although in bad cases ascites not unfre-quently occurs, while there may also be transudation into the pleural and pericardial cavities. The degree assumed by the general anasarca is sometimes enormous, resulting in the ex-tremest distortions. The eyes may be actually closed by the swelling, and movement of the lower limbs rendered almost impossible. Dropsy does not always follow the order here named. Much depends upon the position of the patient.

Thus if he be upon his feet, the latter may be the first to swell, or if he be lying in the recumbent position, the back may be the seat of the first swelling.

Simultaneously with, and sometimes earlier than the drop- sical symptoms, is a *diminution in the quantity and alteration in the quality of the urine.* The former may amount to actual suppression. The latter is manifested by alteration in color, specific gravity, sediment, and chemical composition.

First, as to *color*, instead of the normal amber or lemon-yel- low hue, the urine presents, if acid in reaction, as it always is in this disease when freshly passed, a peculiar *smoke-hue*, due to the presence of a small amount of blood. This peculiar hue is difficult of further description, although once recognized is always remembered. As stated, it requires an acid reaction for its production, and the same urine alkalized presents a brighter red color. The smoke-hue also becomes red if the quantity of blood is large, which is not often the case; but here again the peculiar tint returns if the blood is allowed to sub- side. If the urine is very small in quantity, and there is much sediment, the former is turbid, and may have a brownish tinge. The bloody urine of acute Bright's disease disappears sooner or later before the other acute symptoms subside, and may reappear.

The *specific gravity* of the urine at first is high, 1025 to 1030, partly due to the diminished quantity and partly to the admixture of blood. Later, if the symptoms abate, the specific gravity diminishes with the increase in the quantity. Or if the disease lasts for any length of time, or passes over into the chronic form, a similar reduction in weight occurs; this may result in a specific gravity of as low as 1010.

As to *chemical composition*, the chief alterations is in the *presence of albumen.* This is generally large, the urine often solidifying on the application of heat and acid, while it con- stantly contains more than half its bulk. This albumen is derived in part from the extravasated blood, and in part is a result of the inflammatory action.

I have already referred to the carelessness of expression often

used in indicating the quantity of albumen. Thus it is said
that albumen is 50 or 75 per cent. when this proportion of the
bulk tested is intended. In point of fact it is rare that more
than *a half of one* per cent. of albumen is found in acute nephritis when accurately determined by gravimetric measures.
These are rarely used on account of the time required for their
use. Pains should, therefore, be taken to indicate exactly what is
intended. The actual amounts of albumen found in acute nephritis, though relatively large, may be as low as .2 of one per
cent. Bartels reports a case in which the albumen reached
1.525 per cent., or 12.962 grams (200 grains) in twenty-four
hours.

Next in importance is the reduction in the amount of *urea*,
which is invariable until convalescence sets in. The degree of
diminution is, on the other hand, quite variable. There is a
good deal of range, within the limits of health, in the quantity
of urea eliminated in twenty-four hours—from 20 to 40 grams
may be admitted in the adult. In a case reported by Dickinson the amount declined to .72 grams (11.1 grains) in twenty-
four hours; in another, by Rosenstein, to 1.4 grams (21.5
grains); in another, by Bartels, to .8 of one per cent. (normal
.015 to .03). More frequently the amount is reduced to one-
fourth or one-half the normal. Such diminution, if large, is
of the gravest importance, as some of the most serious symptoms depend upon it.

In general it may be said of the other normal constituents
of the urine, uric acid, phosphoric and sulphuric acid, chlorine,
etc., that they are all diminished, but no clinical significance
attaches to such diminution.

As to *sediment*, the urine of all cases of acute parenchymatous nephritis deposits a sediment which in the early stages,
at least, is copious and brownish or reddish-brown in hue;
later it may diminish in amount and assume a lighter color.
Microscopical examination reveals this deposit to be made up
mainly of casts of the uriniferous tubules, free cells from these
same tubules, blood-corpuscles, red and colorless, and very
constantly crystals of uric acid, together with granular urates.

The casts include the varieties known as epithelial casts, blood casts, hyaline casts, and dark granular casts. The hyaline casts are probably pure. fibrin. The epithelial consist of the same material, to which epithelial cells of the tubules and blood-corpuscles have become attached. The epithelium thus attached, as well as that which is found free in the urine, is variously altered. Some of the cells are merely the seat of cloudy swelling, others are decidedly granular, while others again are converted into compound granule-cells, or exudation cells, by complete fatty degeneration. The latter indicate such extreme derangement of nutrition from pressure, remoteness of the cell from its blood supply, or other cause of loss of balance between the *nutriens* and *nutriendum*, that fatty metamorphosis occurs. Casts containing a few oil-drops may also be present; but much oil is not found until the case has continued for some time, in fact become chronic.

Acute parenchymatous inflammation of the kidneys is rarely, if ever, so intense that suppuration results ; so that pus is almost never found in the urine from this cause. There may be an increased number of leucocytes which have passed through the capillary walls into the tubules during the inflammatory process, and thence down the tubules with such urine as may be secreted, but the number is rarely so great as to constitute pus. I, at least, have never known it to occur. The blood found in the urine doubtless comes from the ruptured bloodvessels of the Malpighian glomerule.

Along with the diminished quantity of urine is often met a disposition to frequent micturition, the efforts at which are only partially successful, resulting in the emission of from a few drops to a tablespoonful. This frequent desire to pass water is a purely reflex symptom, the bladder being free from disease. It sometimes precedes, in point of time, all other symptoms.

The *train of nervous symptoms* usually known as *uræmic*, and ascribed to the accumulation of excrementitious substances in the blood, does not invariably present itself. When present, however, it adds a phase of gravity which dare not be overlooked. We should, therefore, be always on the alert to recognize them.

The first of these usually observed is *drowsiness*, which may be sudden or gradual in its onset, and may be slight or profound. From the latter degree the transition is easy to the next symptom, that of *coma*, which is indeed nothing but a profound sleep from which the patient may or may not be temporarily aroused. Alternating with the latter may be epileptoid convulsions, which are the most alarming and dangerous symptom of Bright's disease. This is not always the succession of these symptoms. Convulsions most frequently succeed drowsiness, but as often precede coma, and they may occur without being preceded by drowsiness while the child is playing and to the unsuspecting parents convalescing. Indeed there may be no suspicion of Bright's disease whatever until convulsions suddenly seize the child and possibly continue with alternating coma, and no return to consciousness until death closes the scene. Drowsiness, in like manner, may be the first symptom of the renal disease to attract attention, others being overlooked or possibly even absent. The convulsions exhibit every grade of movement, from the slightest twitching to the most violent epileptiform spasm.

Impairment of vision or *actual* blindness suddenly occurring is another symptom of acute uræmia, which sometimes supervenes upon other symptoms of uræmia, or it may itself usher in the unfortunate complication. This blindness, it must be remembered, is something altogether different from that which is the result of organic retinal changes, which are rare in acute nephritis, but common in some of the chronic forms. The exact cause of this blindness, in acute uræmia, is as yet undetermined.

Headache and *irritability* of temper are occasionally due to uræmia, while *nausea* and *vomiting* are not infrequent. They show themselves at different stages and in different degrees. When vomiting accompanies scanty or suppressed urine, the vomited matters sometimes exhibit a urinous odor, the elements of urine, more particularly urea and its derivative, ammonium carbonate, being thus supplementally eliminated.

Itching of the skin is another symptom sometimes present in uræmia. It is probably due to the irritant action of the urea upon the nerves of the skin as it is being supplementally eliminated by that organ. That such increased elimination takes place is attested by that rare, but still unquestioned occurrence in which the entire integument is covered with a frost-like coating which has been found upon analysis to be crystals of pure urea.

Another rare symptom of uræmia, which belongs rather to the uræmia of chronic renal disease, is *asthma, uræmic asthma* as it is called. This is a true asthma, a spasmodic contraction of the bronchial tubules, accompanied by the usual labored breathing, which occurs in paroxysms, most frequently at night, and which break up with the appearance of copious frothy secretion just as in the ordinary spasmodic asthma. The view that these attacks are uræmic in origin is sustained by Bartels,* Dickinson,† and Allbutt,‡ but denied by Rosenstein. They are to be distinguished from paroxysms of dyspnœa due to œdema of the lung, which is not infrequent in acute nephritis, by the absence of the fine, moist râles which attend the latter; and from other conditions involving the lungs and pleura, by the absence of the physical signs present in these conditions. Of course it is not impossible for nephritis to happen to an asthmatic whose attacks would then occur as before, independent of the uræmic cause, or might be increased in frequency or rendered more unmanageable by the latter.

The symptoms of acute parenchymatous nephritis which have been detailed are those which may be considered most essential to its recognition. There are, however, others which are less peculiar to it, or occur also in other diseases. These should be referred to in order to give completeness to our picture.

---

* Bartels, op. citat., p. 111.
† Dickinson, op. citat., p. 177.
‡ Allbutt, British Med. Journ., September, 1877.

One of these is *fever*.   It would be expected that in inflammation of the kidney, as in inflammation of any organ, there would be fever.   Such is the case.   But two circumstances combine to make this symptom of little diagnostic value.   In the first place, it is of itself not very marked, and in the second there is very constantly fever in scarlatina to which it is so often superadded, and the increment thus resulting is not sufficient to demark it from such variations in temperature as the original disease itself is subject to.   Of course, the fever of an acute idiopathic nephritis, resulting from exposure to cold, would be more promptly noticed, but here, again, there is nothing distinctive about it.

Pain over the region of the kidney is another of these symptoms.   It is more frequently absent than present, and when present it is not great.   More frequently it may be elicited by strong pressure.   When present it may radiate from its central seat to the groin and surface of the thigh.

Again, acute nephritis, after scarlatina, may be ushered in by *vomiting*, which is probably the result of a reflex irritation of the stomach, and is quite independent of that which has been referred to as the result of uræmia.   Obstinate *derangement of digestion* may result from the same cause, which is particularly manifested when an attempt is made to use solid food.

On the other hand, some of the best-known and most constant symptoms are sometimes wanting.   This is the case with dropsy, which is occasionally absent, especially in cases following diphtheria.   But still more remarkable are those very rare cases in which albuminuria and casts, one or both, are wanting.   This unquestionably happens in grave cases, for short periods of time, but has never been known to exist throughout a case where examinations have been repeated and continued.   Bartels, during an epidemic which prevailed in 1853–54, *met a few cases in which dropsy set in after scarlet fever, although the urine passed by the patients contained no albumen.*   In all these cases very little urine was secreted, in one, only two tablespoonfuls in the twenty-four hours; *very* soon *afterwards a more abundant excretion of bloody urine set in*, while the dropsy increased and the cases passed through the course of scarlatinal nephritis. Henoch reports a case, which is quoted by Bartels, wherein a boy of twelve, three weeks after having had scarlatina, was admitted to the Charité, in Berlin, with œdema of the face and scrotum. His urine was scanty, acid, deposited a sediment, but was *free from albumen*. The microscope "revealed no elements which would indicate the existence of a nephritis," the sediment consisting entirely of amorphous urates which disappeared on the application of heat. During the next two days the œdema increased, but the urine remained unchanged. At the end of this time, at night, he had a violent convulsion followed by unconsciousness. The next day the urine, drawn off by catheter, contained a large amount of albumen and hyaline casts beset with fat-granules. He died two days later and the autopsy revealed well-marked parenchymatous nephritis in both kidneys. In another case of anasarca after scarlet fever, described by Henoch, the urine for two weeks sometimes contained albumen and sometimes not. The only explanation of these cases is that suggested by Bartels, that in the spread of the nephritis over the organ, portions become involved to the extent of actual suppression of urine while others remaining healthy secrete the only urine which enters the bladder. Of course such a thing would be impossible if the entire area of both kidneys is involved, which ordinarily happens sooner or later. Here we must have either entire suppression or albuminous urine. It is possible also that in acute nephritis we may have, temporarily, albuminuria without casts, and casts without albuminuria, but such conditions must be very temporary and exceptional; and the statement of M. Phillippe, referred to by Dickinson as quoted by Jaccoud, that of sixty patients affected with scarlatinal dropsy, there was not one in whom there was albuminuria is simply absurd. I have already shown (p. 59) how in convalescence the albumen may disappear before the casts contained by the tubules had all escaped. On the other hand, the proportion of casts varies greatly;

generally quite numerous, they are sometimes scanty. Albumen, although also varying in amount, may generally be said to be abundant, notwithstanding the small gravimetric percentage alluded to.

There are no symptoms by which the alterations in the Malpighian corpuscles which have been described under the term glomerulo-nephritis are recognizable. We only know that they are most frequently found in those dying of acute nephritis, and that in such cases suppression of urine is a constant symptom, while the occurrence of this symptom is well explained by the histological changes which the Malpighian body has suffered.

The duration of acute nephritis is very variable—from a few days to many months and even years. The former class of cases are usually fatal, for very few which recover do so in a few days. The most rapid usually require a month. As to the cases of longer duration, the possibility of recovery at any time cannot be denied, but nothing is better determined than that the longer the duration the more difficult the cure. Of course such cases cannot be spoken of as acute. An interesting question presents itself: Where shall we draw the line of demarcation between acute and chronic parenchymatous nephritis? A good histological landmark would seem to be the superaddition to parenchymatous changes those of the interstitial tissue, that is, the presence of intertubular nucleation. But this, too, may occur at a very variable period. It has been observed as early as at ten weeks, but I should not call a case chronic that had lasted for this length of time. Perhaps six months might be considered a suitable period, as by this time intertubular changes are probably established in all cases.

Complications.

These are not numerous in acute as compared with chronic Bright's disease, and some which are described as complications are not really such, but local symptoms. Thus, *œdema of the lungs* occurs as a part of the general tendency to dropsy, and may be a grave symptom resulting in death by suffocation. It is not the result of an intercurrent bronchitis. *Pneumonia*, on the other hand, is an occasional legitimate complication. *Inflammation of the serous membranes* is more truly a complication, but not every case in which there is effusion into a serous cavity is of such a nature. Such are, again, local dropsies. But inflammation of serous membranes is rather more prone to occur, while the exudate is apt to assume a purulent character, thus, also, increasing the gravity of the case—indeed causing its fatal termination. Pleurisy is the most frequent form of this inflammation, pericarditis next, and peritonitis next.

Hypertrophy of the heart is not a frequent complication of acute nephritis. It is a well-recognized one of chronic Bright's disease, and is ascribed by some to the impurity of the blood and consequent resistance of the arterioles to its admission into them, and by others to the resistance to the movement of the blood through the diseased kidney. In either event *time* is an essential condition to its production. It is not, therefore, until the kidney disease has existed for some time that it can ordinarily occur. Hence in true acute nephritis it would not be present. If, however, the disease be much prolonged, so as to become subacute or chronic, it may present itself. It occasionally happens, however, that the hypertrophy occurs earlier. Thus Dickinson reports a case recognized at eight weeks, and confirmed by post-mortem examination at ten weeks, in a child of seven years. In children the heart hypertrophies more rapidly than in adults.

Allusion has been made to the gastric symptoms which very commonly attend acute nephritis, especially after scarlet fever. Dr. Fenwick[*] and Dr. Wilson Fox[†] have shown that these are associated with organic changes in the structure of the stomach. Dr. Fenwick characterizes these as gastritis, as evidenced by increased vascularity of the mucous membrane,

---

[*] Fenwick, Samuel, The Morbid States of the Stomach and Duodenum, 1868, p. 177.

[†] Fox, Wilson, Medico-Chirurg. Transac., vol. xli, p. 361.

distension of the tubes by a confused mass of cells and granular matter, and occasional thickening of the basement membrane. To these Dr. Fox has added thickening of the intertubular tissue.

Notwithstanding the frequency of convulsions in acute nephritis, structural alterations in the brain are almost unknown. Apoplectic effusions do not occur in most cases, probably because of the comparative structural integrity of the blood-vessels of the brain in the young, in whom the disease mainly occurs.

In like manner the blindness which not infrequently occurs as a symptom of uræmia, is unattended by retinal changes, nor does that condition known as albuminuric retinitis occur in acute parenchymatous nephritis except with the extremest rarity.

Diagnosis.

The diagnosis of acute parenchymatous nephritis is ordinarily quite easy. The previous history, the usually easily recognizable cause, the suddenness of the attack, the scanty and bloody urine with its high specific gravity, the copious albuminuria, the blood and epithelial and dark granular casts, the blood-corpuscles, free epithelium and compound granule-cells in the urine,—these are a combination of symptoms which admit of only one interpretation. At a later stage, the absence of one or more of these symptoms may somewhat increase the difficulty, but it is scarcely possible to err if those which remain are duly considered. It must be remembered also that an acute condition, such as this described, may supervene upon any one of the chronic forms of Bright's disease to be described.

It is very desirable for the sake of treatment that the renal complication caused by scarlet fever should be recognized at the earliest possible moment. To this end the test of Dr. Mahomed for minute traces of hæmoglobin which are present in the pre-albuminuric stage of scarlatina will be found invaluable, provided his views are sustained. According to him, the first result of the high vascular tension which forms part of the nephritic process, is the transudation of a minute trace of hæmoglobin, before the albumen makes its appearance. Indeed the test is inoperative when albumen is present in the urine. Dr. Stevenson's modification of the test,* acknowledged by Mahomed to be the more brilliant in its results, is performed as follows: To a drop or two of urine in a small test-tube, add one drop of the tincture of guaiacum and a few drops of ozonized ether; agitate, and allow the ether to collect at the top. If hæmoglobin be present, the ether carries with it the blue color that is produced, leaving the urine colorless below. The discovery thus made, according to Dr. Mahomed, often enables the practitioner to avert the inflammation by a brisk purge or copious sweat. The test will also respond just after the albumen has disappeared, as well as just before it shows itself. Saliva, nasal mucus, iodine in minute quantities, all strike a blue color with tincture of guaiacum, some without and some with ozonic ether. Their presence should therefore be guarded against.

Prognosis.

Grave as is justly considered this disease, recoveries from it are numerous, and the prognosis is generally favorable. Even without treatment cases are known to have recovered, and much may be accomplished by a judicious treatment. The prognosis should, however, always be guarded, as insidious causes may produce death when it is least expected. Among the most important of these is uræmia, and I can best illustrate my meaning by narrating a case:

A child about five years old had an imperfectly developed attack of scarlatina, which was considered simple sore throat, and he was allowed to go out of the house in winter weather. Dropsy supervened, and the mother carried him to a homœo-

---

* Dr. Mahomed's original test may be found in the author's Manual on the Practical Examination of Urine, 3d ed., 1880, p. 73.

path, who failed to appreciate the condition or its cause. He prescribed, however, directing the child to return. He did so several times, growing constantly more œdematous. I was finally asked to visit him, when I found enormous œdema throughout the body. A little examination satisfied me as to its cause. The smoky urine was highly albuminous, almost solidified on use of heat and nitric acid, and contained blood- and epithelial casts, with numerous blood-corpuscles. He was placed on appropriate treatment, when the dropsy and albumen diminished. He was a wilful boy, and his indulgent parents again allowed him to be exposed to cold. The dropsy increased, and the albumen as well. So great was the œdema of the prepuce that his urine had to be drawn, and so difficult the introduction of the catheter that it was finally allowed to remain. The scrotum was also enormously swollen, and sloughing ensued. Under appropriate treatment, however, all these symptoms subsided; the catheter was no longer required, the albumen rapidly diminished, and the quantity of urine was sufficiently abundant. I saw him at 2 P.M. of a Saturday, and believed him to be convalescent. Between five and six of the same evening, after slight vomiting, he became suddenly unconscious. I did not reach him for several hours; but all efforts were unavailing, and he died at one o'clock following. Very careful questioning of the mother elicited that he seemed a *little drowsy* on the previous day, but was particularly bright on the afternoon in which coma supervened.

Bartels says that death from urœmia has never occurred in his experience, except when the disease has resulted from scarlatina or diphtheria; but Dickinson narrates a fatal case resulting from exposure, in which death was preceded by coma and other symptoms of evident urœmic origin.*

Pulmonary œdema is another of these causes of sudden death, the patient drowning as it were in his own secretions. Its onset is characterized by shortness of breath, frothy expec-

---

* Op. citat., p. 75-78.

toration, and abundant small râles. *Purulent exudation* into the serous cavities may also precipitate death.

The symptoms of gravest import are therefore those of uræmia, manifested in any one or all the various ways, the presence of any of the complications alluded to, and especially suppression of urine. Cases should not, however, be despaired of even when there is complete suppression of urine. My friend Dr. Wharton Sinkler reports a case of recovery in which there was suppression of urine for five days, following scarlatina.* Always, however, suppression of urine is the gravest of symptoms, and death generally ensues within a couple of days after it sets in. The possibility of sudden death should always be borne in mind, and mentioned to the relatives of the patient, although the number of cases in which this occurs is not very great. Of course the longer the duration of the case, the less the likelihood of recovery.

Treatment.

There is no doubt that many cases of acute nephritis recover while the conditions of *rest, quietude* and *warmth* are maintained. And it is further certain that, whatever other means of treatment are used, these three conditions are absolutely necessary to recovery. A patient with acute Bright's disease, therefore, whatever its mode of origin, should be put to bed, kept quiet, and warmly covered.

I should seldom, however, be satisfied with this mode of treatment alone. The selection of other remedies will depend somewhat upon the severity of the case. If the urine be suppressed, *dry cups*, or in severe cases, *cut cups* to the loins, will so divert the blood as to permit a relief to the stagnation which always exists in the acutely inflamed kidney. These cups should be followed by a warm, moist *poultice* to the same region, which, indeed, should be used under any circumstances, whether the cupping is necessary or not. Poultices

* American Supplement to the Obstetrical Journal of Great Britain and Ireland, Dec., 1878.

should, therefore, always be resorted to, and if the symptoms are at all severe,—that is, where there is complete or almost total suppression of urine, nausea, headache, or delirium,—should be preceded by cupping. Although at first thought it would seem that the kidneys are quite remote from the seat whence the blood is immediately removed, it must be remembered that we are relieving the blood pressure in the lumbar arteries which come off from the aorta near the renal arteries, and thus divert the blood from the latter. Under all ordinary circumstances, dry-cupping is sufficient; cut-cupping should be reserved for the most extreme symptoms, where the strength of the patient has not been previously reduced. Some care must, however, be exercised in the use of dry-cupping, lest we defeat its end. The object of dry-cupping, as justly observed by Dr. G. Johnson, is to facilitate the movement of the blood through the capillaries into the veins,—to draw the blood rapidly through the part, and thus relieve the pressure of the blood in the renal vessels. To do this, the cups must be removed as soon as there is a decided redness, and placed on another part in the vicinity. By allowing them to remain too long, the blood is stagnated in the capillaries, its onward movement prevented, and there is, therefore, no derivation of blood from the involved organ.

The above means have for their object the direct relief of the congestion of the kidney. But this is not the only indication while the kidney is congested. The congestion, in some instances, is altogether due to an excess of work thrown upon it in consequence of suppressed or deficient action of the skin, and in all cases the carrying out of the natural function of the organ tends to increase any existing congestion. Can the kidney be in any way relieved of this functional irritation? Are there any organs, in other words, which can supplement the kidney? The bowels and skin are such organs. A second indication, therefore, is to increase the action of these.

First the bowels. *Purgatives* are useful in aiding the relief of congestion as well as to secure a complemental act of secretion, and for these purposes they should always be employed.

But the reason for which I early employ a purgative is not more for either of these objects than to promote the action of other remedies, a purpose which applies not only to the treatment of Bright's disease, but also to that of all diseases. It is a well-known fact in the absorption of fluids, borne out by the phenomena of osmosis, that this does not take place rapidly when the bloodvessels are congested and there is a slowly-moving current. The well-known experiment of Magendie beautifully illustrates this. This consisted in injecting into the peritoneal cavity a colored fluid, which at first was not appreciably absorbed, but which, on opening a bloodvessel, disappeared rapidly before his eyes. The treatment of any case of acute Bright's disease is therefore well commenced by the use of a cathartic, and after its effect the prompt action of other remedies may be looked for. Indeed, it is almost useless to administer any remedies before some action is obtained from the bowels, as they will be many hours in producing their effects; whereas after such influence they will be as many minutes. For these purposes an aperient should be given early in ordinary cases of acute Bright's disease. The purgative most suitable for this is a saline. A simple dose of bitartrate or sulphate of potassium, of magnesia for children, or citrate of magnesium, or epsom salts for adults will be sufficient. The indication is to get a watery stool as soon as possible. In view of the fact that the stomach is often sensitive, it is desirable to use an aperient which is not nauseous or irritating, and to this end, some one of the delicate effervescing preparations so common in modern pharmacy may be used.

Next, or simultaneously, we have to promote the action of the skin. This is favored by the maintenance of warmth and avoidance of cold, already insisted upon. But we are not confined to these protecting measures. The skin may be made to do the work of the kidney itself, and thus one of the most alarming dangers of Bright's disease, uræmic intoxication, averted, while at the same time the congestion of the kidney is also relieved.

The class of remedies which produce this action are *diapho-*

retics; and, of the simple remedies, none is better than the ordinary sweet spirit of nitre, especially if it be combined with neutral mixture and small doses of ipecacuanha. If more active measures are required, some one of the preparations of *jaborandi* may be used; the dose varying with the effect it is desired to obtain. If moderate diaphoresis only is desired, doses of from ten to fifteen drops of the fluid extract may be given to adults every two hours, and increased, if necessary, until the effect is brought about. To children, five to ten drops may be given in the same manner. The further use of this important remedy, and its active principle, *pilocarpin,* will be again referred to in treating the effects of uræmia.

Another method of accomplishing the same end is by *warm baths,* or, better still, by the so-called warm or cold pack, in which the patient is wrapped in a wet sheet and then enveloped in a sufficient number of blankets. Perspiration is thus copiously induced, and when thus caused, is agreeable, and never attended by the faintness which sometimes follows the use of the hot-air bath,—another means of accomplishing the same end, which will be further considered under the treatment of chronic Bright's disease. In an ordinary severe case of acute Bright's disease, a single pack of this kind will remove all symptoms which may cause anxiety, and happily inaugurate the convalescence, while it may be repeated daily, if necessary.

Nothing has been yet said of the use of *diuretics,* which are, perhaps, the first means thought of by most practitioners in the treatment of Bright's disease, acute or chronic, and, no doubt, in many cases they deserve an early consideration. Yet the propriety of their use has been much disputed, and at first thought there would seem to be legitimate objection to them in the treatment of acute nephritis; for with the idea of increased secretion of urine is generally associated that of an increased flow of blood to the kidney. And the question naturally arises, shall a kidney already congested and inflamed be further jeopardized by crowding more blood into it?

On the other hand, it is well known that convalescence in a

case of acute Bright's disease which has been left to recover without treatment is always ushered in by a most copious diuresis. This is usually explained by the fact that urea itself is a decided diuretic, as may be shown by injecting it into the bloodvessels of any animal,—an operation which is followed by copious diuresis. In the early stages of Bright's disease the urea and other organic constituents are retained in the blood, and when the circulation through the kidney becomes free, they exert their diuretic action. It will be observed, however, that this takes place only after the circulation becomes free, and it must be looked upon, therefore, not so much as a cause, as a result of an improvement in the condition of the organ. Nevertheless, to facilitate such a condition of affairs as copious secretion of urine, and with it the elimination of those effete matters the accumulation of which constitutes the chief danger of Bright's disease,—uræmia,—can only be considered desirable if it can be done without exciting congestion of the kidney.

The secret in the proper use of diuretics lies in the selection of such as effect their object without producing a stagnation; and such there are. To understand this properly, it must be recalled that the secretion of urine is largely a process of filtration, a process of squeezing out the water and dissolved elements by pressure from behind, and that this is accomplished in the Malpighian bodies by the agency of the arterial pressure and the force of contraction of the heart. It must be remembered that there are two sides to the renal capillary circulation, an *arterial* side and a *venous* side. The first consists in the afferent arteriole, and the capillary ball contained in the dilated end of the convoluted tubule, and forming with the latter the Malpighian body; the second, of the capillary network formed by the splitting up of the efferent vessel after it leaves the Malpighian capsule, and closely embraces the convoluted tubules. The area of this is great, and the movement of the blood slow. As a consequence, a condition favorable to increasing the blood-pressure in the Malpighian body exists. Such pressure is obtained by increasing the force of the heart's contraction, or increasing the arterial pressure by the introduction of fluids within the bloodvessels. The effect of this is to produce a more rapid filtration; that is, more water is squeezed out from the bloodvessels into the Malpighian capsules, whence it is carried downward in the tubules. Now, whatever remedies increase the force of the heart's action, or the arterial pressure by absorption of fluids, will increase the amount of water thus filtered out. Such remedies are *digitalis*, the *salines*, and *diluent drinks* generally,—digitalis by increasing the force of the heart's action, the salines and diluents by increasing blood-pressure through their absorption.

Digitalis is certainly the diuretic most to be relied upon, and when combined with the salines, freely diluted, affords a powerful lever for good. It is necessary, however, to have a reliable preparation, and unless one is sure of the quality of the tincture, it is best to use a freshly prepared infusion. At the same time it is also true that much smaller doses of the tincture are usually given than of the infusion. Thus, of the latter, f℥ss. is often administered, equivalent to three and three-quarter grains, while eight minims or sixteen drops of the tincture, equivalent to one grain of the powder, are considered a full dose, a discrepancy which must account for at least a portion of the diminished effect of the tincture. Digitalis should therefore be given in sufficient quantity,—f℥ss. to f℥i of the infusion to children, and f℥ij to f℥ss. to adults,—repeated every three hours until an appreciable effect is produced on the rate of the pulse, when it should be diminished. Not until then can we look for a diuretic action. Digitalis, when thus administered, should, of course, be watched, and the patient should be seen twice a day until an effect is produced. I prefer to give it at first alone. Of the salines with which it may be combined, acetate, citrate, and bitartrate of potassium are to be preferred. Their diuretic action doubtless depends upon the impetus they give to the osmosis of fluids which hold them in solution, thus increasing the arterial tension and contributing to the flushing of the kidney. To adults half a drachm of these may be given every two or three hours freely diluted, because water itself is an excellent diuretic; ten grains to children as often. There can be no doubt that an increased filtration of water into the Malpighian capsules aids the separation of the organic constituents in the second capillary network referred to, both by facilitating osmosis, and by washing out from the secreting cells of the convoluted tubules the organic matter already excreted by them.

Another admirable diuretic combination, including all of these elements, is Trousseau's diuretic wine, which consists of:

Junip. contus,	.	.	.	.	.	.	.	℥x
Pulv. digitalis, .	.	.	.	.	.	.	.	ℨij
Pulv. scillæ,	.	.	.	.	.	.	.	ℨj
Vin. xerici,	.	.	.	.	.	.	.	Oj.

Macerate for four days and add

Potas. acetatis, .	.	.	.	.	.	.	℥iij.

Express and filter.

S. Tablespoonful three times a day for an adult.

By such means as these, after the unloading of the blood-vessels by the action of a purge, we may greatly serve our patient by diuretics.

On the other hand, turpentine, cantharides, copaiba, and the class of diuretics which produce a congestion and stagnation of blood in the second or venous capillary network, are mischievous, and should not be employed.

Infusion of digitalis may also be used in the shape of fomentations. Cloths wrung out in hot infusion of digitalis and laid over the abdomen of the patient, have been known to produce diuresis when all other measures have failed. But this is not likely to be called for in mild cases.

The *diet* of patients with acute Bright's disease, while it should be nutritious, should be of the simplest and most easily digestible character. The irritability of the stomach in this disease has been alluded to, and it is important that it should not be excited. Milk may be considered the typical food, not merely because of its easy assimilation and nutritious character, but because there is abundant testimony to prove that albuminuria diminishes under its use, while the proportion of urea contributed to the blood is less than by animal flesh. While solid animal food is not to be recommended, there is no reason why animal broths and beef teas should be precluded, provided it is desirable to break up the monotony of a milk diet. Rice and farinaceous preparations generally, are suitable adjuvants to the milk diet. The combination of lime-water, and still better of carbonated water with milk, should not be overlooked in the treatment of the sensitive stomach.

Treatment of Acute Uræmia.—The alarming and dangerous character of the symptoms of this condition lead me to a separate consideration of the measures required in their treatment. The treatment which has just been described is such as would be called for by an ordinary case of acute nephritis of a decided character. The tendency of it will be to prevent the retention of those effete matters, whatever their precise nature, which constitute the essence of uræmia, while it is not intended to reduce the patient by secreting and purging. But notwithstanding all of our efforts in this direction they sometimes fail, and we are called upon to contend against convulsion or coma, or more frequently both in alternation. How shall they be met? The indication has already been explained. It is the retention of effete urea and its allies which causes the uræmia. These must therefore be gotten rid of. The kidneys are not acting and the secretion of urine is suppressed. There remain therefore but the bowels and skin to operate upon. But the patient is unconscious and cannot swallow voluntarily. Such remedies must therefore be used as do not require his co-operation. These are *croton oil* and *elaterium.* Of the former two drops, slightly diluted with plain oil or glycerin, or in case of extreme necessity undiluted, may be introduced into the mouth, whence it is quickly absorbed. Its operation may be facilitated by a rectal injection. Of elaterium a quarter of a grain in powder may be introduced into the mouth.

In like manner the skin may be made to substitute the action of the kidney. At the present day *jaborandi* is the most efficient agent by which to accomplish this. The most convenient method is by subcutaneous injection of its active principle *pilocarpin.* For an adult one-third of a grain of the dissolved muriate thus administered, is generally sufficient to excite the most profuse diuresis and salivation within half an hour. Should it not, the dose may be repeated. In the absence of pilocarpin the freshly prepared infusion of jaborandi may be injected into the rectum, with almost equally prompt results. Care must be taken to keep the bulk within limits lest it be rejected. Four ounces of hot water may be poured upon a drachm of jaborandi leaves, and when sufficiently cool strained and injected. The doses here referred to are intended for adults. My friend Dr. Horace Williams has used the fluid extract in suppository, with excellent results. A fluid drachm may be inspissated and put into a single suppository. They should be reduced in size, for children, suitably to the age.

I have never seen any of the extreme prostration which is said sometimes to result from the use of jaborandi. There is undoubtedly a feeling of weakness and relaxation after a copious sweat from this as from other causes, which may be combated by stimulating and supporting measures where they can be applied, but under the urgent circumstances here supposed this needs not be considered. Dysuria is said to occur occasionally, resulting from jaborandi.

Should these measures produce their physiological results of purgation and sweating, without relief to the uræmic symptoms, they may be repeated in four to six hours.

The *hot-air* bath, *warm-water* bath, and *warm pack* which were our sole resources for these purposes before jaborandi and its preparations came into use, may still be used if the latter fail or are not at hand. The hot-air bath is easily used. A tin pipe two or three inches in diameter with an expanded extremity under which a spirit-lamp is placed, while the other end is placed under the bed clothing, will answer the purpose very well. An ordinary rain-spout may be used. The warm pack or bath is not so available in acute uræmia because of the unconsciousness of the patient and his inability to help himself. Sometimes even the hot-air bath fails to make them perspire, and all efforts seem unavailing.

Under these circumstances I should not hesitate to take a small quantity of *blood from the arm* for the relief of uræmic convulsions and coma, provided the patient is not too feeble. No one doubts the efficiency of bleeding in puerperal convulsions, and if puerperal convulsions are uræmic as I believe they mainly are, then bleeding should be of service in the uræmic convulsions of acute Bright's disease. My friend, Dr. Hiram Corson, of Conshohocken, Pa., is the only physician I have known to bleed in the convulsions of acute nephritis after scarlatina, but he informs me that he has done so with great advantage. I wish to be clearly understood, I do not advocate bleeding for the cure of the nephritis. It is simply suggested as a rational measure for the relief of the convulsions when other measures fail.

The hydrate of chloral should not be forgotten; indeed it is one of the most valuable remedies for convulsion, and should be one of the first measures tried. In the case of an adult, a drachm may be injected into the rectum in solution; fifteen to thirty grains for a child. Its use is sometimes followed by the promptest favorable results.

The use of *opium* requires to be alluded to. The caution which has always been suggested in its use I believe to be, in the main, a wholesome one, and I should prefer to produce hypnotic, sedative, and antispasmodic effects by chloral and the bromides whenever it is possible. I am sure I have seen death accelerated in one case of previously unsuspected chronic Bright's disease, in which large doses of opium were exhibited for another purpose,—overdoses, in fact, but quite insufficient of themselves to produce the fatal result, which was preceded by uræmic stupor. After death the urine was drawn by a catheter and found to be albuminous, and a post-mortem examination revealed a contracted kidney. It is well known that Professor Loomis, of New York, treats with apparent success cases of uræmic convulsions with hypodermic injections of large doses of morphia (one-half grain or more), doses which I would fear to use under ordinary circumstances in the absence of renal disease.

Treatment of Complications.—Complications should be treated by remedies called for by such conditions independent of the renal cause. Effusions into the pleural cavities and abdomen are often best relieved by paracentesis, or aspiration; pneumonia and bronchitis by counter-irritation.

These same measures which have been detailed, excepting the general bloodletting and chloral, may also be employed in the treatment of suppression of urine or of obstinate dropsy without uræmic symptoms, with such modifications as circumstances may suggest, due regard being paid to the strength of the patient. They will be further referred to when discussing the treatment of the chronic forms of Bright's disease.

Sooner or later also, in the treatment of acute parenchymatous nephritis, supporting treatment is rendered necessary to repair the losses which the blood suffers by the albuminuria, and to some extent also by the depleting measures of treatment. These effects should indeed be anticipated by proper diet, tonics, quinine, especially iron, wine, malt liquors, whiskey, or brandy as indicated. These measures will also be more particularly alluded to in the treatment of chronic Bright's disease.

SECTION VI.

CHRONIC PARENCHYMATOUS NEPHRITIS.

Synonyms.—Chronic tubal nephritis; chronic catarrhal nephritis; chronic diffuse nephritis; large white kidneys.

CHRONIC PARENCHYMATOUS NEPHRITIS IS A CHRONIC HYPERPLASTIC PROCESS IN THE KIDNEY, WHICH HAS ITS ESSENTIAL AND PRIMARY SEAT IN THE EPITHELIUM OF THE TUBULES, BUT IN WHICH INTERSTITIAL CHANGES OF THE NATURE OF NUCLEAR PROLIFERATION ARE ALSO MORE OR LESS CONSTANTLY PRESENT.

Etiology.

The etiology of chronic parenchymatous nephritis cannot always be traced. While it is frequently a simple continuation of a process which begins as acute parenchymatous nephritis, more frequently it originates *de novo*. To the former category, *scarlatina* and *pregnancy* contribute the greater number. Of a certain number of the second category, the causes are undiscoverable. *Habitual exposure* to *cold* and *moisture* doubtless produces some cases. Residence in damp houses may thus cause it. *Chronic suppuration*, such as occurs in bone disease, phthisis, tertiary syphilis, psoas abscess, etc., are among the causes of this, as well as of the lardaceous form of chronic Bright's disease.

Long-continued exposure to *malaria* is now a recognized cause of chronic parenchymatous nephritis. Indeed, Bartels says that next to chronic suppuration it is the most frequent cause, but I have never myself been able to trace a case to this cause. Dr. S. C. Busey, of Washington, D. C., presented a paper at the session of the American Medical Association, in 1880, on "Malaria as a cause of Bright's Disease in Children." T. Koike published a case of nephritis from malaria, in the Tokei *Medical Journal*, March 27th, 1880.

Although *alcohol* is not a common cause of chronic parenchymatous nephritis, yet I cannot but think that the chronic nephritis which we find in confirmed drunkards, those who are always saturated with whiskey when they can get it, owes its presence to the latter agent. To be sure it cannot be denied that the exposure to which these outcasts are subjected, may be the cause.

That *mercury* is a common cause of chronic nephritis, as was believed by the older physicians, is denied by Bartels, who bases his denial upon the most extended experience in the use of mercury for syphilis in the hospital at Kiel.

In all the modes of origin of chronic nephritis the rationale would seem to be that some noxious agent in the blood, whether introduced from without or retained there from deficient action of the skin, produces the alteration in the kidney by a slow but constant irritation of the epithelium which attempts to remove it.

Morbid Anatomy.

There are two distinct stages in the morbid anatomy of chronic parenchymatous nephritis, if the disease is of sufficient duration, viz., the stage of enlargement, or the *large white kidney*, and that of contraction, or the *fatty and contracting kidney*.

I. *Stage of Enlargement.*—There are few more striking objects in morbid anatomy than a typical example of the *large white kidney*, as the product of this stage of chronic parenchymatous nephritis is called. The kidney is large, smooth, white, or slightly tinged with yellow; weighs generally from seven to ten ounces, but is often much heavier. It is usually doughy, sometimes elastic in consistence.

The capsule, which may be thinner than in health, strips off easily, but occasionally drags a little of the parenchyma with it. When the smooth white surface thus uncovered is examined, the little capillary circlets bounding the lobules in the normal organ are in some places indistinct, in others conspicuous; the same is true of the stellate veins of Verheyn. Numerous yellow specks are seen scattered over the surface. Hæmorrhagic extravasations are also occasionally present, but very much more rarely than in the acute form. Alongside of these the greater translucency of more nearly normal areas results also in a mottled hue.

On section it is at once evident that the enlargement resides altogether in the *cortex*, which is at the same time markedly anæmic, its intense white contrasting strongly with the pink hue of the cones, which, though paler than in health, are much less so than the cortex. Closer examination of the cortex reveals the same yellow specks found on the external surface. They contribute, with similar alterations less decided, to form a series of dull white striæ which alternate with somewhat broader translucent striæ radiating towards the surface; the former correspond to the area of the convoluted tubules and Malpighian bodies,—the labyrinth,—the latter to that of the medullary rays. In consequence of extreme swelling of the masses of convoluted tubes which dip down between the cones, as seen in longitudinal section, there may be produced the same sheaf-like distortion of the latter which was referred to as sometimes occurring in acute parenchymatous nephritis.

The pelvis of the kidneys in chronic parenchymatous nephritis is the seat of catarrhal swelling and a slight degree of hyperæmia.

Minute Change.—Microscopic examination of thin sections shows the involvement of both *tubes and intertubular substance.* Turning our attention first to the former, many are found choked with granular cells and the granular debris of cells, causing them to appear, under the microscope, as black opaque lines by transmitted light, very similar, indeed, to the tubes in acute nephritis. In other situations the tubules are filled with fat-globules and cells in a state of fatty degeneration. Again, the two elements are combined. In places the lumen of the tubes is preserved, in others not. In other situations the cells are nearly normal. The parts presenting a yellow tinge are those in which the fatty elements have replaced the normal, and this is the composition of the yellow specks already alluded to as visible to the naked eye. They represent a coil of tubules filled with oil-drops or fatty cells.*

Here, again, certain tubules contain casts, usually of the waxy

---

* Very great differences are noted in different kidneys in the amount of oil present in the cells, which have never been satisfactorily explained. Dickinson says the cells have a greater tendency to be fatty when cold is the cause.

kind. Sometimes, indeed, they are very numerous. Rarely hæmorrhagic extravasations are also found in the tubules.

The *capillaries* of the cortex are completely or nearly empty of blood, which has been expressed from them by the distended tubules. To this, it need hardly be said, is due the extreme whiteness of these kidneys, whence the name by which they are known. The Malpighian capillaries are subject to the changes already described, of proliferation of capillary nuclei; so, too, the alterations of the epithelium of the glomeruli there described may also be present as well as those of the capsular epithelium. From the first would result thickening and more or less opacity of the capillary walls. The muscular walls of the afferent arterioles are sometimes hypertrophied, and the vessels dilated in consequence of the resistance to the entrance of blood into the glomerule, but general arterial hypertrophy is not constant in this stage of the disease. The intertubular capillaries and veins present no changes except those already referred to as the result of the compression by the distended tubes.

The *medullary cones* in chronic parenchymatous nephritis are more altered than in the acute form, but the changes in them are quite secondary in importance, and their microscopic appearance is scarcely altered. They are sometimes a little paler, owing partially to a granular and fatty alteration in the cells lining the straight tubules, and partly to the presence of similar cells pushed down from the convoluted tubules above them. On the other hand, they may even be slightly deeper in hue from congestion. The straight tubes of the cones as well as the looped tubes of Henle often contain waxy casts.

In chronic parenchymatous nephritis the *interstitial tissue* is always altered, and, it may be said, proportionately to the duration of the disease. It has already been said that any case of parenchymatous nephritis, sufficiently prolonged, is attended by a hyperplasia of connective tissue cells, although it is difficult to say when this overgrowth begins. Langhans reports* a case in which death, occurring five weeks after the appearance of the first symptoms directly

---

* Loc. citat., p. 105.

traceable to a thorough wetting, the stroma was *markedly thickened*. And in a case of Dickinson's already alluded to, intertubular cellular formation, "though approximating as much to pus as to fibre," was found within six weeks of the outset. Again, cases of much longer duration may be entirely without it. Interstitial fibrosis may, however, be considered as a superaddition of chronicity, and wherever a case is distinctly chronic it may be inferred, with tolerable certainty, that it is present.

The interstitial change is generally conceded to be the result of the long-continued hyperæmia. But it will have been observed that it does not usually make its appearance until the primary hyperæmia begins to decline, and although the latter may contribute, in a degree, yet I am inclined to think it is rather the result of a secondary hyperæmia, induced by the irritation of the cellular accumulations in the tubules. These act as foreign bodies and induce interstitial hyperplasia, just as the cheesy collections excite interstitial hyperplasia in the lungs, where catarrhal products in the bronchial tubules are always, sooner or later, attended by interstitial change.

In this overgrowth the thickness of the trabeculæ of tissue between the tubules varies extremely, being sometimes so slight as to be discoverable only on microscopic examination of thin sections; at others it is appreciable to the naked eye.

Minute examination reveals the thickened trabeculæ to consist of numerous round and oval nuclei, between which may be a homogeneous or more or less distinctly fibrillated intercellular substance.

II. *The Stage of Atrophy—The Fatty and Contracting Kidney, or the Large Contracting Kidney.*—The interstitial new formation above referred to possesses all the characteristics of new connective tissue formed elsewhere. It invariably exhibits a tendency to contract, and in doing so, it gradually distorts the shape of the previously enlarged organ, while it obliterates also a variable quantity of its tubular structure. The degree of this distortion varies greatly, increasing with the duration of the process. The kidney continues as large and even larger than the normal organ.   It is, however, smaller than the large white kidney, uneven, lobulated, but never presents the hobnailed or granular appearance of the kidney of interstitial nephritis.   Its capsule does not strip off easily as from the large smooth organ, but drags with it considerable of the tubular structure.   The capsule removed, however, the surface of the kidney exhibits, between the constrictions, the same pallid speckled appearance, distinct stellate veins, etc., already described; and on section the same enlarged anæmic cortex.

Microscopically, sections exhibit the same alternation of groups of normal and choked tubules already described, alongside of other places in which the tubules together with the Malpighian bodies at their extremities are obliterated.   Between them is found a large amount of interstitial tissue, and the Malpighian bodies are surrounded by concentric layers of the same.   Even minute cysts, the result of obstruction of tubules by the constricting tissue, are found.   These will be more fully described when we come to consider the contracted kidney of interstitial nephritis, with which also this stage of atrophy in chronic parenchymatous nephritis will be more carefully contrasted.

It not infrequently happens that along with the changes constituting chronic parenchymatous nephritis, are found, to a certain extent also, those of lardaceous disease.   Thus in a large white kidney, the Malpighian bodies will often strike the mahogany red reaction with iodine, characteristic of this condition, although the alteration may not be recognizable by the naked eye.   Occasionally the alteration may even affect the afferent and efferent vessels.   This secondary amyloid change is ascribed, by Dr. George Johnson, to the exhaustive drain which is constituted by the long-continued albuminuria, such drains, as will be more fully considered under lardaceous disease, being the common cause of this disease of the kidney.

9

Symptoms of Chronic Parenchymatous Nephritis.

There are few distinctive symptoms of chronic parenchymatous nephritis, but by a thorough investigation of the case in all its bearings a diagnosis may generally be made.

Dropsy is, however, almost always present, and it is very apt to be general, at least involving the subcutaneous connective tissue generally—the face, hands, feet, legs, thighs, and trunk. The serous sacs also frequently contain fluid, almost always in severe cases. But dropsy is not always thus general. It may be confined to the extremities or to the face, and in a case now under my observation, where the diagnosis is peculiarly sustained by the characteristics of the urine, the dropsy is confined to the scrotum, which is enormously swollen, while there is no œdema elsewhere. It is not impossible that dropsy may be entirely wanting, but as a rule no symptom is more constant, and none gives the patient so much inconvenience, as this one of dropsical swelling. His legs and thighs are twice their normal dimensions. They are so heavy he cannot lift them, while they are often excoriated, and moist with exuding serum, and smarting with irritation. Very frequently, as the result of spontaneous rupture of the skin, the discharge of serum is profuse, saturating the bedclothing, and even dropping upon the floor; occasionally, also, with relief to the patient.

Another very constant symptom is *anæmia*, producing a peculiar translucent waxy appearance, which is really very characteristic when present in marked degree, and is often alone sufficient to suggest the disease. But there may be very slight degrees of it which are not at all peculiar.

Again, the *debility* of those suffering with this condition is very striking. If able to walk at all, they soon get out of breath, are immediately exhausted. Locomotion is often impossible in consequence of the extreme swelling, even if the strength otherwise permit it.

Now if such a patient be questioned, and it be found that he took cold three or four months previous, following which immediately was noted this dropsy, which did not disappear, or which disappeared and returned; or if it be found that he had scarlet fever some months or even years previous, following which there had been swelling which had been more or less continuous,—if these points are made out the diagnosis is easy. For it has been a case of acute parenchymatous nephritis which has become chronic. If, however, there be no such history, but an insidious beginning, traceable to any one of the causes named, or to no cause, the case is not so clear. But the urine affords additional symptoms.

Condition of the Urine in Chronic Parenchymatous Nephritis. —The urine is diminished, but somewhat variable in quantity, pale in color, of low specific gravity, highly albuminous, and deposits often, but not always, a copious white sediment.

First, as to *quantity*. This, as stated, is diminished, but quite variable, ranging from 300 to 1200 cc. (10 to 40 oz). It is, however, seldom suppressed, as it sometimes is in acute nephritis. The quantity of urine also increases as the patient improves, or as the stage of contraction is entered upon, so that it may even exceed the normal.

The *specific gravity*, notwithstanding the small quantity of urine, is less than the average of health. It varies somewhat, inversely with the quantity of urine, but the most usual range is from 1008 to 1015.

The *albumen*, while also large, varies as to its percentage amount with the quantity of urine passed—from 1 to 5 per cent., or from one-half to seven-eighths the bulk of the urine tested. The variation in the 24 hours' quantity of urine should always be remembered. This is particularly important in testing the value of various therapeutic measures, some of which, by increasing the quantity of water passed, diminish the percentage of albumen without diminishing the 24 hours' quantity. This is often overlooked. The amount of albumen lost in the urine is sometimes enormous. It has even occurred that the percentage proportion of albumen in the urine has exceeded that in the serum of the blood from the same patient. Bartels accurately determined the 24 hours' quantity in several cases under his observation, and found in one instance an average of 17.36 grams (267.85 grains) daily, for the last month of the patient's life; in another an average of 15.28 grams (235.76 grains) daily, for two months; and in a third 10.04

grams (155 grains), for six months. The last lost 6804 grams, or about 18 pounds troy, in a period of 27 months. From this may be appreciated the enormous drain upon the system by this "hæmorrhage" of albumen. The quantity of albumen has very little effect upon the specific gravity. Indeed, the lighter urines are generally those which have the larger amount of albumen, because highly albuminous urines contain little urea.

The *copious white sediment* is made up of variously granular casts, among which the *dark* granular are conspicuous by their numbers and size, and especially their width. There are also found oil-casts, and casts containing entire and fragmentary epithelial cells, which are also granular and oily. Finally, yellow waxy casts are also found. These, which are also generally of larger diameter, are of especial value in indicating the chronic nature of the disease. Casts generally increase in numbers with the progress of the disease, being at first less numerous. They are not, however, always thus numerous, being indeed sometimes wanting. Much granular debris, similar to that making up the black granular casts, and probably therefore derived from the disintegration of epithelium, is also found free in the urine.

Leucocytes are also often very numerous in the sediment of chronic parenchymatous nephritis, while red corpuscles, although occasionally present, are much less common.

The sediment, when mixed with the urine, gives the latter a turbid, dirty appearance, which is sometimes permanent, but when the sediment has subsided the supernatant fluid is tolerably clear.

The *normal constituents* of the urine are *generally diminished* in quantity. The most important of these is *urea*. To the reduced amount of solids, and particularly of urea, the reduced specific gravity is due. And although the quantity of albumen is large, and must therefore increase somewhat the specific gravity, such increase is not ordinarily sufficient to compensate for the decline.

The above statements with regard to the urine are borne out by my own experience; but Bartels, who has added so much to our knowledge of chronic nephritis, adds some others apparently at. variance with them.   I desire, therefore, to make separate mention of them.   He states that at the height of the disease, when the smallest quantity of urine is secreted, *the specific gravity is regularly above the normal.**   He has frequently found the specific gravity of such urine above 1040, higher, consequently, than the specific gravity of the blood serum before it was secreted.   This was determined by actual trial of the serum of the blood drawn from the arm at the time.   He also says the proportion of albumen to the specific gravity is a pretty constant one, *rising as the specific gravity rises,* and diminishing as the specific gravity falls.†   I have already said that it is the urines of low specific gravity which contain more albumen, the low specific gravity being due to the small quantity of urea which these urines contain.   He adds, however, that the specific gravity of an albuminous urine does not by any means correspond with a certain fixed percentage of albumen, while we are only justified in assuming that the percentage of albumen has increased or diminished, because the specific gravity has risen and fallen, when the change in the latter takes place within a very short period. Further observations may remove the apparent discrepancy.

Notwithstanding the small proportion of urea which is excreted in this affection, *uræmia is rare,* especially so before the stage of contraction is reached.   And if it be asked why this should be, I know no better reason than that assigned by Bartels, that all the physiological sources of urea production are less active.   The appetite and digestion are bad, less food is taken, and thus the principal source of urea in the economy is cut down.   Further, the nitrogenous tissues are wasted and tissue change is less active, so that the urea from these sources is less produced.   Furthermore, some of it may be stored in the abundant serum which occupies the interstices of the tissues and the serous cavities which share in the dropsy.

Are there any symptoms by which we can recognize the stage of contraction, which takes place sooner or later, provided the patient lives?   The most valuable evidence that this has

---

* Op. citat., p. 357.   † Op. citat., p. 360.

occurred is the presence of hypertrophy of the left ventricle. Although the possibility of an earlier hypertrophy in children cannot be denied, it is very seldom that it occurs in parenchymatous nephritis prior to the stage of contraction. From this I would not have it inferred that I believe the hypertrophy is caused by the contraction and the consequent difficulty encountered by the blood in its movement through the kidney. This was Traube's view of the production of hypertrophy of the left ventricle in the contracted kidney of interstitial nephritis, and is held by Bartels and others. I am inclined to believe that the hypertrophy is due rather to the resistance to the movement of the poisoned blood through the arterioles of the entire system, and not the kidneys alone. This was the original view of Bright. Dr. George Johnson has further shown that this resistance results in a thickening of the muscular coat of the bloodvessels, which still further augments their power to resist the heart. This contracts still more powerfully to overcome the increased resistance, and becomes further hypertrophied. As already stated, time is required to reach this stage, and by the time hypertrophy is developed contraction of the kidney is likely to have occurred. Long duration of the disease also affords presumptive evidence that contraction has taken place. This cannot always be ascertained. But if a case come under observation as a case of undoubted parenchymatous nephritis, and continues under observation for a year or more, the process of contraction is likely to have commenced.

The *dropsy* diminishes and may disappear as the stage of contraction is entered upon. So also the urine changes in its properties. The quantity, previously small, is increased, while the specific gravity remains low; the quantity of albumen is also much less than during the stage of inflammation. In these respects—absence of dropsy, large amount of urine, and small amount of albumen—it resembles the true contracted kidney of interstitial nephritis, with which indeed it may be confounded in the absence of a previous history. But the casts continue numerous, and exhibit much the same character that they do in the stage of enlargement, although they too may become scanty; and if we have not a knowledge of previous history the diagnosis between contraction secondary to previous enlargement, and primary contraction the result of interstitial nephritis, may be impossible.

Uræmia is rather more common in the stage of contraction than that of enlargement, but still comparatively infrequent. When present it exhibits the symptoms already detailed under acute parenchymatous nephritis.

The duration of chronic parenchymatous nephritis is very variable. Many cases terminate unfavorably within a year after they have been established; but I have one now under observation in the stage of contraction which I have known to exist for seven years. Another patient has just died who was under my care four years. Both of these cases apparently originated from cold.

Complications.

The complications of chronic parenchymatous nephritis are the same as those of acute. Œdema of the lungs, bronchitis, pneumonia, inflammation of serous membranes are all liable to occur. Hypertrophy of the left ventricle is more common than in acute nephritis, for the reasons already referred to, but still very much less so than in interstitial nephritis. Derangements of digestion are very constant, probably due to a more advanced stage of the structural changes described under acute nephritis. The acute blindness unattended by retinal changes, described as occurring in the uræmia of acute nephritis, rarely occurs here, while retinal changes are rather more frequent, but still uncommon compared with interstitial nephritis, under which they will be described.

Diagnosis.

The diagnosis of the inflammatory stage or stage of enlargement is ordinarily easy. The extreme pallor of the patient, the diminished urine of low specific gravity, the very large amounts of albumen, the numerous dark granular, oil- and waxy casts of large diameter, free fatty cells and compound granule cells, especially if we are able to trace a history of long duration, all point to the disease. And if there is an antecedent history of scarlatina or exposure to cold, pregnancy or long exposure, there can be no mistake.

The symptoms of amyloid or lardaceous kidney very closely resemble those of the large white kidney, and it has been mentioned that the same causes are capable of developing both. Occasionally it is absolutely impossible to say which form of disease is present. It has usually been considered that if there is enlargement of the liver and spleen, or persistent diarrhœa, and the cause is one which may produce lardaceous disease, it is certain that the latter condition exists; but recent observation has shown that the first two at least may be present, together with all the causes and other symptoms which are regarded as favoring lardaceous disease, and yet the disease be parenchymatous nephritis.* As a rule there is not so much dropsy in lardaceous disease, casts are more scanty, and generally hyaline, though sometimes oil-casts are found. Sometimes, too, the two forms of disease coexist, either as the result of the same cause, or the amyloid disease may be the result of long-continued parenchymatous nephritis.

The stage of contraction is more difficult of recognition unless we have had the case for some time under observation, and are able to trace its continuation with the stage of inflammation. The resemblance to the contracted kidney of interstitial nephritis may otherwise be very close. But here again the albuminuria is apt to be larger and the casts more numerous, and to include the numerous varieties mentioned instead of the scanty, small pale granular casts which attend interstitial nephritis. In the latter the quantity of urine exceeds the normal, while in the former, although the quantity is larger than in the stage of enlargement, it is still within the normal.

Prognosis.

The prognosis of chronic parenchymatous nephritis is unfavorable so far as recovery is concerned. Most cases terminate

---

* See an article by Dr. Paul Fürbringer, Zur Diagnose der amyloiden Entartung der Nieren. Virchow's Archiv, Bd. 71, 1877, s. 400.

unfavorably within two years after they are thoroughly recognized, and sometimes within a few months. Many cases, however, may be very much prolonged by treatment, and if prolonged to the stage of contraction the patient may be tolerably comfortable for some time, seems indeed to have another lease upon life. But sooner or later the dropsy returns, and the patient dies of exhaustion. Or some one of the complications, or possibly uræmia, intervenes to carry him off. Of the former, œdema of the lungs or of the glottis, and pneumonia, are particularly dangerous. In the stage of enlargement, uræmia, while it is of rarer occurrence, is also less apt to end fatally than in the stage of contraction.

Treatment.

While it occasionally happens that spontaneous recoveries from acute nephritis occur, this is far from the case with the chronic form. Here the expectant plan of treatment does not suffice. The patient with chronic parenchymatous nephritis, if left alone, grows steadily worse, and although measures of treatment may not frequently result in recovery, they very often, if judicious, cause marked improvement, and long avert the fatal end.

There is always an intermediate stage between that of acute nephritis and the condition of the large white kidney, from which recovery often takes place, which calls for a modification of, or an addition to, the treatment described for the acute, and which is indicated by an impaired quality of the blood, due partly to the gradual accumulation of effete matter, and partly to the drain upon the system of the large albuminuria. But, as it is a condition growing out of the prolonged presence of the disease, it is practically covered in the treatment of the chronic form, and requires, therefore, not to be separated from it.

The chief indications in the treatment of chronic parenchymatous nephritis are two: *first* to improve the quality of the blood, which has become anæmic and loaded with urea and allied organic compounds; and, *second*, to combat the symptoms and complications which form a source of great inconvenience and danger to the patient.

The first of these indications is chiefly fulfilled by the use of iron, salts of quinia and strychnia, nourishing food and proper hygienic influences; and also by depurating the blood of its retained urea. It may be laid down as a rule, to which there are few exceptions, that the *continuous* use of some one of the preparations of *iron* is indicated. The well-known Basham's mixture, really a solution of acetate of iron, and made by adding to tincture of the chloride of iron, acetic acid and solution of the acetate of ammonia, has the advantage of at least tending to eliminate, while it also strengthens.* But the tincture of the chloride of iron, alone, is a powerful agent, which is always accessible, and when combined with the sweet spirit of nitre, is perhaps as efficient as the Basham's mixture. To either, the quinia and strychnia salts may be added, if desired; while to the latter the infusion or tincture of quassia makes a compatible addition.

With regard to *food*, while it is true that an abundance, and of good quality is desired, a question has properly arisen as to the propriety of using the highly nitrogenized substances, as animal flesh. It is now well determined that the urea formed in the blood and eliminated in the kidneys is derived chiefly from the azotized elements of the food, and that the more nitrogenous food we consume the more urea accumulates in the blood, which the kidneys in their disabled condition are unable to remove. The results of reasoning have been sustained by experience, and it has been observed that where the appetite is good and large quantities of meat are eaten, uræmic convulsions have been more frequent, whereas, when the appetite has been bad, and little food taken, uræmic convulsions in chronic nephritis are very rare. While, therefore, it is not desirable to omit all such food, it *is* desirable to limit it to mod-

* The formula for Basham's mixture, which I commonly write, is as follows: R. Tinct. ferr. chlorid., f3ij; Ac. acet. destillat., f3ij; Liq. ammon. acetatis, f3iij; Curaçoe or syrupi simpl., Aquæ, ãã q. s. ad f3vi: M. et S. Tablespoonful three times a day, in half a tumbler of water.

eration, and, while drawing elements of mixed food from the
vegetable kingdom, to make up the deficiency in meats by the
free use of milk. The good results of the milk treatment in
cases of chronic nephritis are now among the best acknowledged in the treatment of the malady, as evidenced in the diminution of albuminuria, decline in dropsy, increase in the quantity of urine passed, and general amelioration of symptoms.
And of the different methods of employing milk, the pure milk
diet has been most satisfactory. I find that from three to
three and a half quarts a day meet the requirements of an
adult male, this quantity having been used for months by one
of my patients, without any practical change in weight. The
milk should not be skimmed, for it is by retaining the cream,
that the nitrogenous principle, the casein, is maintained disproportionately small. The better way is to drink a fixed
quantity at stated intervals, say, six to eight ounces every two
hours. As to the rationale of its operation, there is reason
to believe that it operates by affording an easily assimilable
food *freely diluted*, which can be taken in sufficient quantity to
provide the forces of the economy without surcharging the
blood with nitrogen. When the idiosyncrasy of the patient is
such that a pure milk diet cannot be borne, it should constitute
as large a proportion as possible.

An invariable effect of the milk diet in my experience is an
increase in the quantity of urine passed, which would of course
result in a diminished percentage of albumen, while the absolute quantity in twenty-four hours might be unchanged. But
in my observations I have avoided this source of error, and
succeeded in demonstrating an absolute diminution in the
quantity of albumen passed.

Next to milk, *rest* is the most useful measure to ameliorate
the symptoms of chronic nephritis. I have frequently, by
means of the milk treatment, reduced the albuminuria to a
point beyond which, however, it seemed impossible to influence
it further in walking cases of chronic nephritis. The same
case has then been put to bed, the milk diet continued, and a
still further reduction of albuminuria has immediately ensued.

A reversal of the method has been as promptly followed by the opposite condition. The patient was allowed to get up, the milk diet being still continued, and immediately the albumen increased. He was then allowed to use a mixed diet, and as promptly a further increase in the albumen appeared. The beneficial effect of rest upon œdema from any cause is too well recognized to require other than an allusion.

The advantages of rest in bed are sometimes more than counterbalanced by the disadvantage to the patient of confinement and want of fresh air and outdoor life. These of course must be weighed, and that one adopted which serves the patient best.

Under hygienic measures is included *suitable clothing*. That next the body should be of wool; for it must be remembered that, on the one hand, the skin is a powerful adjuvant to the kidney in its eliminating operations, and, on the other hand, that any interference with or suppression of the action of the skin must throw more work on the kidney. Cold is the agent which produces such suppression, and warmth the means by which the action of the skin is encouraged; and no texture prevents the former or secures the latter more effectually than wool.

For the same reason, while the maximum amount of fresh air is desirable, *cold* and *dampness* should be avoided or sufficiently guarded against.

The second indication is to depurate the blood of accumulated impurities, as well as to combat the symptoms and complications which cause inconvenience or jeopardize life. These symptoms are those of dropsy, effusions into the serous cavities, and congestions. The patients suffering from them are usually confined to the house, or go out of it at such great inconvenience as to make it intolerable to do so. Of dropsy there is abundant evidence to the naked eye; but of the necessity of depuration there is, unfortunately, no direct means of estimation except by a volumetric analysis of urine, which involves so much trouble and care as scarcely to be possible to the general practitioner. Fortunately, however, the means which are best calculated to relieve the one are most likely to relieve the other. These measures are, in addition to diuretics, such as promote a more decided action of the skin than any yet alluded to, and certain purgatives.

With regard to diuretics, nothing need be added to what has been already said, bearing in mind that digitalis is our most powerful lever. But with regard to measures which promote a decided action of the skin, I desire to add a little more. These are the *warm bath, warm-pack bath*, and the *hot-air bath* already alluded to. The latter, in consequence of its more ready application, is to be preferred whenever it can be borne. Previous to the introduction of jaborandi, I used a great deal in my wards at the Philadelphia Hospital, the hot-air bath, and made some observations to determine its value. The results of these satisfied me that we have a much more useful agent than many of us have suspected. A patient with large white kidney was under my observation for more than a year. During a portion of this time his urine was carefully measured, and a portion of the twenty-four hours' urine analyzed for urea by Liebig's volumetric process, which was repeated to insure accuracy. He was a very large man, passing copiously of urine, and the quantity thus arrived at was 35 grams (540 grains); the total quantity of urine being 2000 cubic centimeters (66⅔ f℥). He was then ordered a hot-air bath daily, during which he perspired most freely. The twenty-four hours' urine was of course diminished; but on estimating the urea in the twenty-four hours after the sweating had been continued three days, it was found to be 46.3 grams (714 grains) in 1700 cubic centimeters (56⅔ f℥) urine,—actually an increase over the amount secreted when not using the baths. This can be accounted for by the increased celerity of the circulation which would naturally result. If we add to this the amount of urea contained in the increased perspiration, which was of course not determined on account of the difficulties of collection, we will perceive how powerful a means of depurating the blood of its urea is thus at our disposal; and I am quite certain that if the use of the hot-air bath were more common, our power over

Bright's disease would be greater. There is a common impression that it is troublesome and difficult of application. But this is not the case with the simple appliance described when discussing the treatment of acute nephritis. Still simpler is it to place the patient upon a chair having a perforated seat, placing a spirit-lamp under the latter, and covering the patient and chair by an india-rubber waterproof. Sometimes, however, these hot-air baths are not well borne by patients; they do not perspire, the head and face become flushed, and the former throbs and aches. The latter symptoms may often be relieved by tying a wet handkerchief about the head, as is done in the Turkish bath. In the event of failure, however, the "warm" or "cold-pack" may be used, or the warm bath. In the former the patient is wrapped up in a wet sheet, either warm or cold, and further enveloped in a sufficient number of blankets. A very comfortable sweat generally ensues. In the use of the warm bath the patient is immersed in it at a temperature of about 40° C. (104° F.), and kept there from half an hour to an hour. He is then removed and wrapped in blankets.

These measures may be resorted to on alternate days, or for a short time daily. It may be objected that they are exhausting to the strength of the patient; but I think they will be found less so than is commonly supposed. Due regard should, of course, be paid to this tendency, and the strength of the patient may at the same time be maintained by alcohol, milk, iron, and tonics.

It has been said under the treatment of acute parenchymatous nephritis that these effects are more conveniently and as efficiently brought about by the use of *jaborandi* and its derivative *pilocarpin*. The directions for their use, given in connection with acute nephritis (p. 116), need not be repeated here. They may be used about as often as the baths, usually on alternate days, occasionally daily, with advantage.

The use of *purgatives* for depurative purposes and to reduce the dropsy has long been common in the treatment of chronic nephritis, and to this end it has been common to select a peculiar class of purgatives, viz., those which produce profuse watery evacuations, as elaterium, scammony, gamboge, and jalap.   In addition to the indications to relieve general venous congestion with a view to promoting absorption, the advantage to be derived from the use of a brisk, prompt cathartic has already been alluded to in speaking of the treatment of acute Bright's disease.   But it must be remembered that in the circumstances now under consideration it is not a temporary cause the effects of which we desire to obviate, but a constantly acting one, so that to be of service the purgative must be continued day after day, or every other day at least.   Now, such use of the hydragogue cathartics above mentioned cannot be continued for any length of time without materially reducing the strength of the patient, much more decidedly than by the daily sweat.   I do not deny their effect in diminishing the dropsy.   On the other hand, I have many times observed this effect, and in some I have observed the dropsy totally disappear,—but with it the strength of the patient to such an extent that as the dropsy subsided the life of the patient went out with it.   I am not, therefore, very partial to the continued use of cathartics in chronic Bright's disease.   But it is to the prolonged use that I refer.   To relieve a sudden emergency, as the occurrence of uræmic symptoms,—in a word, under the same circumstances under which I would use them in *acute* Bright's disease if they could be administered,—would I give them. So, too, it may be sometimes of advantage to alternate their use with the sweat treatment referred to.

Of the remedies mentioned, undoubtedly the one which most strikingly produces the desired effect is elaterium. The profuse painless discharges which it effects in doses of one-tenth to one-fifth of a grain are well known, while the small quantity required makes it peculiarly easy of administration.

But in most cases of chronic nephritis, a stage is finally reached at which all treatment of the kind described fails to relieve the dropsy, which becomes eventually the sorest burden of the malady. The body becomes greatly increased in weight, the integument of the extremities is stretched almost to burst-

ing, and sometimes it does rupture, when it is attended by a
leakage, which, although in one way inconvenient, is in many
senses a great relief to the patient, by diminishing the tension
referred to.   Acting upon this, physicians have long been in
the habit of puncturing the swollen parts to produce the required leakage.   In my early experience I once had such
horrible results in the sloughing away of the entire scrotum of
a little child with scarlatinal nephritis, after I had punctured
it, that I declared I would never repeat it.   But as other
cases came under my observation my prejudices thus excited
gradually disappeared, and I now resort to puncture when it
seems likely to give relief.   It only remains to determine the
best method of performing the operation.   It is a common
practice to make a number of minute punctures with a needle
or sharp-pointed bistoury.   Dr. George Johnson, of London,
recommends making a free incision half an inch long, just
above the outer or inner ankle of each leg, and deep enough to
enter the areolar tissue beneath the skin.   This may be done
with a bistoury; but Dr. Johnson used an instrument mounted
like a spring-lancet, which he recommends as more efficient
and less painful than the repeated fine punctures.   He relates
an instance which is so remarkable and so admirably illustrates the possibility of recovery when the symptoms have
reached an advanced stage, that it is quite worthy of re-
narration.   In July, 1861, he saw a clerk, aged 22, who had
suffered from general dropsy since the end of March, after
exposure to cold.   The urine became nearly solid on addition of
acid and heat, while it *contained numerous oily casts*.   Purgatives and diuretics failed to lessen the dropsy, and at the beginning of September the swelling was so great that the skin
cracked and water oozed through the fissures.   The legs were
now incised; a copious discharge of water occurred, and the
urine became more copious.   From that time he steadily improved; the dropsy passed away, and gradually the urine
ceased to be albuminous; but it was not until the end of
April, 1862, more than a year from his illness, that all traces
of albumen had disappeared.   The chief medicinal treatment

after the incision of the legs was the use of tincture of per-
chloride of iron three times a day, and a dose of broom tea in
the morning. Such recoveries as this are rare, while their
possibility shows the value of hopeful perseverance in treatment. I have never seen the instrument referred to, but
have made the large incisions with satisfactory results, although I can point to none so satisfactory as Dr. Johnson's.
Dr. Dickinson,* on the other hand, relates a case in which, for
the relief of dropsy, one leg was punctured by a needle, and
on the other a lancet was used. He does not say how deep
were the incisions by the lancet, or whether they were mere
punctures; but the openings made by the needle healed without any bad result, while those made by the lancet gave rise
to deep suppurations, pus being discharged through five of
the punctures.

The treatment of the complications is in no way different
from that of the same conditions under other circumstances.
The point to be impressed is the importance of being constantly
on the lookout for them. *Œdema of the glottis* requires sepa-
ate allusion as a complication most alarming and threatening
to life. Inhalations of steam may be tried, but prompt punctures or incisions are the only certain means of relieving the
patient and saving his life.

Special Methods of Treatment.—The above described principles of treatment are those which, modified by the peculiar
requirements of each case, are usually found most satisfactory.
With regard to special curative measures directed to producing
structural change of the kidney and a return to its normal
histology, or directly diminishing albuminuria, I have not had
very satisfactory results, and most of the measures which have
been from time to time suggested are completely useless.

Calomel has been used for long periods to the production of
its specific effects. For what object, except to hasten the blood
dyscrasia which is the ultimate cause of death in this form of
nephritis, I do not know. It requires to be mentioned only to
be deprecated.

_____

* Op. cit., p. 67.

Tannic acid and *gallic* acid I have tried systematically without any result in diminishing the albuminuria.

Ergot in some one of its forms I have thought had some effect in diminishing albuminuria, but in no degree comparable to the effect it produces in hæmorrhage from the kidney, where it is undoubtedly useful. The most convenient form of administration is a pill of ergotin containing three grains, of which two or three may be given at a dose. The fluid extract of ergot may be given instead, in the dose of half a drachm to a drachm three times a day, but is not so well borne by the stomach.*

The fluid extract of *eucalyptus globulus* I tried in a single case, in doses of ten drops three times a day, without any results. This remedy is strongly recommended by Drs. J. B. Leary and William Anderson,† of Brooklyn, New York, and deserves a further trial. Dr. Anderson uses it in connection with the milk diet, to which some of the effect must be attributed.

Several years ago it occurred to me that sandal-wood oil, so efficient in catarrhal conditions of the bladder, ought also to be of service in catarrhal nephritis, and I began to administer it in the form of capsules to patients with albuminuria, giving at first three a day, and increasing to six a day, two after each meal. Its use was followed by a fall in the amount of albumen in some cases, but those intractable cases which refused to yield to other remedies, failed to respond also to the sandal-wood oil. But there is reason why sandal-wood oil should do good, and I hope others will give it a trial in cases of obstinate albuminuria. Dr. Edw. T. Bruen, informs me that he has seen albumen diminish under its use.

The most recent remedies suggested for albuminuria are

---

* The preparations of ergotin made by different manufacturers are probably not of the same strength. That of McKesson and Robbins, of New York, purports to be such that 1 grain equals 10 grains ergot, a 3-grain pill being equal to half a teaspoonful of the fluid extract.

† Proceedings of the Medical Society of the County of Kings, N. Y., vol. iv, No. 6, August, 1879.

fuchsin and *rosanilin*. Feltz and Bouchet* were apparently the first to use them in pills and mixture in doses of three grains a day. They say that under their use, albumen soon disappeared from the urine, and that both these coloring agents are relatively harmless and well borne by the organism. Most recently,† Professor E. di Renzi, of Genoa, has published the results of treatment by fuchsin. He reports a decided fall in the quantity of albumen under its use, as well as of mucus, which he says is often present in the urine of Bright's disease. He ordered it in solution or in pills, preferably in the latter shape, 2½ centigrams (3.8 grs.) to a pill, beginning with 5 centigrams (7.7 grs.) and increasing it to 25 (38.5 grs.) in the twenty-four hours. The remedy produced a marked coloration of the urine. If it does not pass over into the urine it is useless in diminishing the albuminuria. I have not as yet had an opportunity of testing this new remedy.

My friend Dr. Albert H. Smith has used with apparently decided advantage, in two cases, the *benzoate of lime*, in doses of 10 grains every three hours. I failed to obtain the same results in a single case, but am anxious to repeat the trial.

Professor di Renzi also made use of *apomorphin* in doses of 5 to 6 centigrams (7.7 to 9.2 grs.) a day without any effect whatever upon the albuminura.

He found *rest in bed* a very useful measure, *and when united with the milk diet* found it more effectual than any other measure in reducing the albuminuria.

I have already expressed my preference for these latter measures with a view to diminishing the albuminuria. Should they fail, others of course should be tried. For the reduction of the dropsy the systematic diaphoretic treatment, preferably by jaborandi, but also by the hot air and warm bath, are most useful, having due regard to the strength of the patient, which must be kept up by iron, tonics, and, if necessary, alcohol in the shape of the malt liquors, wine, or even the strongest alcoholic preparations.

---

* Deutsch. Med. Wochenschr., 1879.
† Virchow's Archiv, vol. 80, p. 510, June, 1880.

SECTION VII.

LARDACEOUS DISEASE OF THE KIDNEY.

Synonyms.—Amyloid disease, albuminoid disease, waxy kidney, depurative disease.

LARDACEOUS DISEASE IS THAT FORM OF CHRONIC BRIGHT'S DISEASE IN WHICH THE TISSUE ELEMENTS OF THE KIDNEY ARE MORE OR LESS INFILTRATED WITH A PECULIAR ALBUMINOID SUBSTANCE, RESEMBLING MOLTEN WAX OR BOILED STARCH, BUT WHICH IS BEST RECOGNIZED BY ITS STRIKING A DEEP MAHOGANY-RED INSTEAD OF THE ORDINARY YELLOW COLOR, WITH A SOLUTION OF IODINE.*

Etiology.

The most frequent cause of lardaceous disease is acknowledged to be profuse and long-continued suppurative discharge, such as occurs in chronic bone disease, whether syphilitic or traumatic in origin; or such discharge as constitutes the expectoration in cases of chronic phthisis and chronic bronchitis with bronchiectasia. Syphilis itself, independently of the tertiary conditions which it produces, is a frequent cause of lar-

---

* I do not consider it necessary to discuss the nature of the lardaceous or waxy material. Its albuminous composition is now everywhere acknowledged, C 53.58, H 7, N 15.04, O 24.38 being its formula (Kekule and Friedreich), while that of albumen is C 52.7 to 54.5, H 6.9 to 7.3, N 15.4 to 6.5, O 20.9 to 23.5, S .8 to 2 (Gorup-Besanez). The question as to whether the deposit is a metamorphosis or an infiltration is not so definitely settled, but it seems to me to deport itself altogether like the infiltrations, and unlike the metamorphoses; I shall therefore include it among the former until some good reason is found for altering this position. I retain the term *lardaceous disease* because it defines the physical characters of the deposit better than any other, although not the chemical ; further, because it is the term adopted in the revised nomenclature of the Royal College of Physicians of London.

daceous disease.   Chronic dysentery, ulceration of the bowels, chronic albuminuria itself, may all become the causes of this condition.

Age and Sex.—Either sex is equally subject to lardaceous disease, but as men are more frequently exposed to its causes, it is in them rather more common.   Very young children are rarely affected, for evident reasons. Dr. Dickinson has known a case to be fatal at five years of age, and refers to a case of Dr. Gee, in which the lardaceous change was found in the spleen but not in the kidneys of a boy two and a half years of age who had had a profusely discharging abscess of the thigh. The following table from Dickinson shows the distribution as to age of sixty-one cases:

Age.									No. of cases.
0 to 10,	.	.	.	.	.	.	.	.	3
11 to 20,	.	.	.	.	.	.	.	.	11
21 to 30,	.	.	.	.	.	.	.	.	21
31 to 40,	.	.	.	.	.	.	.	.	10
41 to 50,	.	.	.	.	.	.	.	.	10
51 to 60,	.	.	.	.	.	.	.	.	3
61 to 70,	.	.	.	.	.	.	.	.	3
Over 70,	.	.	.	.	.	.	.	.	None.

Morbid Anatomy.

The incipient stages seldom present alterations recognizable by the naked eye unaided by reagents.   But if, after section of the kidney, the cortex be treated by a solution of iodine and iodide of potassium,* numerous mahogany-red points make

---

* *The Test Solutions.*—The best test solution for macroscopic purposes is one made by dissolving .162 grams iodine by the aid of .324 grams of iodide of potassium in 30 cc. of water (2½ grains iodine, 5 grains iodide of potassium to a f℥ of water).   The solution contains about one-half per cent. of iodine.   For microscopic preparations a solution weaker than the above, or a one-fourth per cent. of iodine dissolved by twice the quantity of iodide of potassium, is more suitable, and sometimes a solution containing as much iodine as water alone will take up answers best.   There is a decided difference of opinion as to the effect of the subsequent addition of

their appearance; or if by a solution of violet-anilin, as many red or pink points. These are the Malpighian bodies, whose capillary tufts are the first to be affected by the change. The kidney in this early stage is normal in size, or very slightly

---

sulphuric acid to preparations treated with iodine. Virchow[1] originally announced that a blue or violet color was assumed by the amyloid tissues thus treated. In this he is sustained by all subsequent German pathologists, including the most recent. Most English writers, on the other hand, either deny this reaction altogether, or speak of it as uncertain and unreliable. Thus Dickinson[2] says: "This appears to be an error of observation, partly arising from the fact that when sulphuric acid acts upon iodide of potassium—a salt generally present in the test solutions used—a precipitation of iodine takes place, which usually blackens the tissue." So, also, a Russian writer, Morochowetz,[3] says the reagent is useless because sulphuric acid produces with iodine solutions alone, a violet or blue coloration by a precipitation of iodine crystals. The discrepancy is undoubtedly due to methods of manipulation. Bœttcher[4] and Kyber,[5] whose papers are among the most careful and practical recently issued, both declare that a delicacy is obtained by the use of the sulphuric acid which far exceeds that of any other test. Both ascribe the want of success to the use of too concentrated sulphuric acid. Kyber uses an iodine and iodide of potassium solution of the above strength for macroscopic work, and watery solutions of iodine for the most part for microscopic preparations; and sulphuric acid *diluted* fifteen to twenty times with water. Bœttcher uses the solution of iodine and iodide of potassium, above given, for both macroscopic and microscopic work, and 7 to 8 cc. of sulphuric acid to 100 cc. of water. Kyber also says that in the treatment of sections by the iodine solution, the normal tissues should not be allowed to assume a deeper tint than a pure yellow. If they are *browned* the section is worthless for further treatment, and should be thrown away.

The Anilin-violet Solution.—Jürgens[6] and Heschl[7] in Germany and Cornil[8] in France were the first to call attention to anilin-violet as a test for the lardaceous substance, although the late Dr. Bennett, of Edinburgh, had previously called attention to the fact that carmine and magenta produce distinct coloration. The anilin-violet, of which a 1 per cent. solution is

---

[1] Virchow's Archiv, Bd. vi.

[2] Op. citat., p. 231.

[3] St. Petersburg Med. Weekly, vol. iii, 1878, p. 75.

[4] Bœttcher, Arthur, Beobachtungen über die amyloide Degeneration der Leber, Virchow's Archiv, Bd. 72, 1878, s. 506.

[5] Kyber, Eduard, Weitere Untersuchungen über die amyloide Degeneration, Virchow's Archiv, Bd. 81, 1880, s. 1.

[6] Eine Neue Reaction auf Amyloidkörper, Virchow's Archiv, Bd. 65, 1875, s. 189.

[7] Sitzb. d. Wiener Akad., iii. Abth., Oct., 1876.

[8] Archiv de Physiol. Norm. et Path., t. vii, p. 673, Paris, 1875.

enlarged. Its capsule strips off readily, leaving an organ which exhibits no peculiarities, or a paleness or translucency which readily escapes notice, but may be more easily recognized at the edges of a thin section. Very often too, they are completely overshadowed by other alterations, which are especially apt to be associated with these slight degrees of lardaceous disease. Thus the large white kidney of chronic parenchymatous nephritis may exhibit this degree of lardaceous change, and the latter altogether escape notice without the use of iodine. Hence *the iodine reaction should be tried upon all kidneys* whose morbid anatomy we may be investigating. Under the microscope, however, thin sections exhibit a transparency of the structures involved which does not require iodine to secure its recognition, but beautifully distinct demonstration results from its addition to such preparations.

In a more advanced stage of uncomplicated lardaceous change the kidneys are both enlarged, usually symmetrically, but the extreme degrees of enlargement are usually associated with fatty degeneration of the epithelium. Such organs were a pair weighing 23 ounces, which came under Dr. Dickinson's[*] notice. Dr. Johnson[†] refers to a case in which the two kidneys weighed 28 ounces. Rindfleisch[‡] has seen a single instance of

suitable, strikes a red or pink color with the amyloid material, which contrasts with the violet staining of the normal tissues, and beautiful preparations for the microscope may be thus obtained. Eberth[1] and Fürbringer[2] are recent German writers who prefer this test solution, which the former says is decidedly to be preferred to the iodo-sulphuric acid test to demonstrate the first beginnings of the amyloid change, and Fürbringer also says it is a more delicate test. Kyber says it is this very stage which is best shown by the iodo-sulphuric test. Eberth also says that the anilin-stained microscopic preparations also keep the longest, which is probably true. The blue tint received by the normal tissues, however, rapidly fades, so that the contrast is not maintained.

In my own experience I have found the simple iodine reaction without the use of sulphuric acid sufficiently distinctive for practical purposes.

* Dickinson, op. citat., p. 249.   † Johnson, op. citat., p. 104.

‡ Rindfleisch, Path. Histology, New Syd. Soc. Trans., 1873, vol. ii, p. 167

[1] Eberth, C. J., Die amyloide Entartung, Virchow's Archiv, Bd. 80, 1880, s. 138. An excellent paper.

[2] Fürbringer, Paul, Zur Diagnose der amyloiden Entartung der Nieren, Virchow's Archiv, Bd. 71, 1877, s. 400.

that very rare condition, *complete* lardaceous infiltration, that is, in which the basement membrane of the uriniferous tubes, as well as the capillaries, were infiltrated, where the kidney was enlarged to nearly twice its normal size.

In the uncomplicated forms of lardaceous disease the capsule is not adherent, but if interstitial changes exist to any extent it is adherent. When removed, the surface of the kidney is pale and anæmic; occasionally the stellate veins are conspicuous. The characteristic translucency may even be recognized in the organ in bulk, but in sections is more striking. When the change is present in high degree the edges of a thin section are almost as translucent as a similar section of bacon.

On bisection the cortex is seen to be enlarged; it is pale, anæmic, waxy, firm, and resisting. The pyramids are normal in hue and area. The iodine solution added to such a kidney produces its peculiar coloration, not merely in the Malpighian capillaries, but also in the afferent and efferent vessels and the vasa recta of the pyramids.

In a still later stage, that of *atrophy*, the kidney becomes contracted, diminished in size, rough, and even distorted in shape. The capsule is adherent, and on section the cortex is found narrowed, sometimes as much so as in the contracted kidney of interstitial nephritis.

The most diverse views as to the cause of this *contracted* lardaceous kidney are held by authors. Thus Dr. Grainger Stewart and Dr. Johnson ascribe it to a wasting and destruction of the epithelium and tubules; Virchow, Klebs, and Bartels to a simultaneous or previously existing contracting disease. Dr. Dickinson believes that it is due to the contraction of a new-formed intertubular tissue, which is the result of an irritant property of the lardaceous material, similar to what occurs in the second stage of parenchymatous nephritis. This, once formed, contracts in accordance with its invariable tendency, and produces the rough granular appearance and diminished size of the organ. Rindfleisch also supposes the amyloid change of the vessels to be primary; that the mechanical obstruction thus resulting induces a collateral hyperæmia

of the cortex ; this produces the conditions favorable to a
cellular hyperplasia, which succeeds immediately or upon the
addition of some further irritant.   I do not doubt the correct-
ness of the latter view in the majority of instances, and that
the further irritant is the lardaceous material itself, as Dickinson suggests.   It is of course not impossible for a case of
lardaceous disease to be ingrafted on one of interstitial nephritis.

As to the *microscopy* of this disease, in the first stage, in
which the naked eye often fails to detect anything abnormal
without the aid of iodine, a microscopic examination of thin
sections reveals a lustrous or waxy appearance of the Malpighian bodies, due to an infiltration of their capillary tufts by
the peculiar lardaceous material.   They are also enlarged, in
consequence of a thickening of the capillary walls.   At this
stage there is no visible alteration in the tubules or their epithelium.

In the second stage we have the involvement of larger vessels, the vasa afferentia and efferentia, in the cortex, and also
the vasa recta of the cones.   The other capillaries of the cortex
are also involved, and an exudation occurs into the tubules, of
a glistening material which forms casts.   I have already discussed these casts on p. 68.   Dickinson* believes the material to be, occasionally at least, identical with the lardaceous
substance, although he admits that it very rarely exhibits the
peculiar reaction with iodine.   According to Grainger Stewart,
this material presents neither the peculiar translucency nor
the coloration with iodine which are characteristic of the waxy
degeneration, but exactly resembles the material of hyaline
tube-casts.†   Cornil has never found them to strike the red
reaction with the anilin-violet solution, while Jürgsen obtained
the reaction with this solution in several instances.   The result of my own efforts in treating these casts with iodine after
their extrusion from the tubes into the urine, after many previously ineffectual trials, is shown in the plate opposite p. 68.
It is to be remembered, too, that similar casts are found in the
tubules in other forms of chronic renal disease.

---

* Op. citat., p. 253.   † Op. citat., p. 126.

The *arteriole walls* are thickened by an involvement of both interna and media. This thickening is attended by an extraordinary distinctness of the muscular fibre cells of the circular coat. The middle or muscular coat, since the original announcement by Virchow,[*] has always been considered the first involved. But Cornil[†] has recently asserted that in every case the lesion is localized in the internal coat of the renal vessels. It is true the interna is sometimes disproportionately changed in the small arteries, but this does not necessarily alter the seat of primary deposit. A very distinct demonstration of the thickening of the muscular walls is obtained by injecting the kidney wtth a transparent injecting mass, and then examining sections by the microscope, although the injection of such kidneys is very difficult.

As to the involvement of the *tubules* themselves and their *epithelium* there is some difference of opinion. Dr. Johnson[‡] insists that, in the large majority of instances, the changes in the secreting structures are primary, and says further that he has not met with a single case in which thickening of the bloodvessels in any form was unassociated with extensive changes in the secreting structure of the kidney.

Klebs and Rindfleisch admit that the lardaceous change only rarely attacks the basement membrane and epithelium. The latter says that amyloid infiltration of the tubes is found chiefly in the papillæ, where it is accompanied also by similar changes in the vessels, especially the vasa recta; thence it radiates into the pyramids of Ferrein. He has never seen it in the convoluted tubes, although he does not doubt the possibility of its occurrence there, where it is hard to distinguish the diseased tubes from the vessels which are similarly affected.[§] Grainger Stewart has seen the cells present the swollen, dimly translucent appearance, but never the peculiar coloration. He

---

* Cellular Pathology, Chance's Translation, Philadelphia, 1860, p. 417.
† Archives de Physiol. Normal. et Pathol., t. vii, 1879.
‡ Op. citat., pp. 104–106.
§ Rindfleisch, op. citat., p. 145.

has often found the basement membrane thickened and waxy-looking, without any coloration taking place on the application of iodine, but on a few occasions he has seen that coloration.[*] Bartels[†] says, "a similar change may subsequently affect the tunica propria and epithelium of the tubuli uriniferi." Dickinson describes alterations in the tubules and their lining cells, due to their infiltration with lardaceous material, as the result of which "their normal structure is displayed with abnormal distinctness." . . . . . "Fibrinous casts are abundantly formed and displaced, and yet the epithelial lining of the tube undergoes no disturbance."[‡] Axel Key also admits the involvement of the cells, and holds that the waxy casts in the amyloid kidney are the direct result of the fusion of the epithelial cells which have succumbed to the amyloid infiltration. Bœttcher[§] describes the change in the renal cells as "decided," and Cornil[||] says they sometimes become the seat of amyloid infiltration, when they are transformed into little glassy blocks which exhibit the characteristic coloration with iodine and sulphuric acid.

The facts appear to be these : The bloodvessels are first involved, for the excellent reason that, whatever is the exact morbid condition, the source of the infiltrating material is the blood. Naturally, therefore, the walls of the vessel-walls become first involved. Beyond are the epithelium and basement membrane, which are also capable of the alteration. That the cells of the liver become thus the subject of waxy change all except E. Wagner[¶] admit, and reasonably also may those of the kidney, with the basement membrane on which they are seated. We would expect them, however, to be more recently involved, and the change may be delayed indefinitely. When cells become the subject of the lardaceous change, they acquire

---

* Grainger Stewart, op. citat., p. 127.

† Bartels, op. citat., p. 519.   ‡ Dickinson, op. citat., p. 254.

§ Bœttcher, Virchow's Archiv, Bd. 72, 1878, s. 534.

|| Manuel d'Histologie Pathologique, Paris, 1876, p. 1017 ; American Translation, 1880, p. 46.

¶ Archiv der Heilkunde, Bd. ii, s. 486.

a peculiar translucent glassy hue, are enlarged, and lose all distinctness of outline, neighboring cells appearing to be fused together. The basement membrane also becomes thickened and translucent.

It is also very common for the epithelium of the cells to be in a state of fatty degeneration, and the capillary walls to contain aggregations of fat-globules, while the urine in the latter stages contains oil-casts and fatty cells.

From the above considerations it is evident why the kidney in this stage is enlarged. The arterioles and capillaries are thickened and occupy more bulk, while the cells are swollen, the basement membrane of the tubules thickened, and the tubes themselves more or less distended with the material of the casts.

In the third or contracting stage of lardaceous kidney, minute examination reveals, in addition to the appearances described, the hypernucleated intertubal overgrowth, already referred to as causing by its contraction that of the kidney itself. Cysts are occasionally present for the same reason that they are found in the granular contracted kidney of interstitial nephritis, and an approximate rough granulation is also sometimes assumed, although never the typical granular appearance of interstitial nephritis.

It is not very rare to find at least the Malpighian capsules in the large white kidney the seat of the lardaceous change, but the iodine test is, in my experience, generally necessary to demonstrate it. Dr. Johnson first suggested that it is the direct result of the drain upon the system incident to the large albuminuria which attends parenchymatous nephritis. But Dr. Dickinson* says that, while among his earlier observations he thought he found reason to admit this cause occasionally, later experience has served to show that where the lardaceous condition is associated with one of the other forms of renal disturbance, the former, lardaceous, is usually the primary change. Bartels also calls attention to the very frequent coincidence of

---

* Op. citat., p. 245.

amyloid disease of the kidney with other renal affections, and especially with chronic parenchymatous nephritis in both kidneys, of which, he says, both processes are apparently "co-effects of one and the same cause."

Dr. Dickinson's present views are certainly not sustained by those cases occasionally met, in which are found, first, in a marked degree, the causes which all acknowledge to be most efficient in the production of lardaceous disease; second, copious albuminuria and casts, with extreme dropsy, and, finally, death, with the post-mortem examination revealing the most typical form of large white kidney, in which, however, the iodine solution also reveals amyloid infiltration of the Malpighian tufts only. On the other hand, the fact, which can hardly be controverted, that the same set of causes is capable of producing either disease, makes it much more reasonable to consider, that when both conditions are present, they are "coeffects of one and the same cause," as claimed by Bartels.

Symptoms and Clinical History of Lardaceous Disease.

An individual who has had syphilis, or who has phthisis, chronic bronchitis, bone disease, or other affection in which there is an exhaustive drain, observes that he feels always intensely weary, has no disposition to exertion whatever, or even to rise from his bed. Increased frequency of micturition may be observed at the same time, such that he may have to rise once or twice during the night to micturate. But this symptom may be totally absent. Accidentally, perhaps, a somewhat copious albuminuria is discovered. At first no casts are met, or they are exceedingly scanty, a single one being found in several successive slides, or one or two may be found on a single slide. Those which are noted are hyaline or faintly granular. Later, a slight œdema of the feet may make its appearance while the patient is up and about, but disappears during the night while he is in bed. The albuminuria is now copious but still varies, and casts may be more numerous, or may still be scanty and continue hyaline or faintly granular. The

urine is now decidedly increased in quantity, 1600 to 2500 cc.
(53 to 80 oz.), its specific gravity low, 1005 to 1015.   The
patient exhibits a worn and cachectic appearance, which may
be present earlier, and is sometimes the first symptom which
strikes the attention of the observer.   There is sometimes a
peculiar fetor of the breath.   Still later, all these symptoms increase; the dropsy is persistent, the urine loaded with albumen,
and, in addition to the ordinary delicate hyaline casts, may contain
the glistening waxy casts.   Fatty casts and free fatty epithelium
from the tubules of the kidney may be superadded, as well as
free oil-drops.   Epithelial casts are unfrequent.   Dropsy now
becomes general, involving the arms, trunk, and face, as well
as the lower extremities, and even the serous cavities, the peritoneum, pleura, and occasionally also the pericardium.   Œdema
of the lungs may also occur as a serious complication.   In general, however, it may be stated that dropsy is seldom as extreme as in parenchymatous nephritis.

Towards the close of the disease, the urine, which had been
increased, becomes diminished in quantity, but is seldom suppressed, indeed seldom falls below (600 cc.) 20 ounces.

Of the solid chemical constituents, it may be said of all, that
they are, as a rule, slightly diminished, but not sufficiently to
influence the course of the disease.   It is in consequence of
this that uræmia is almost unknown in lardaceous disease, the
urea and extractives being eliminated in sufficient amount to
avert this evil.

Convulsions or uræmic symptoms of any kind seldom occur,
probably for the reasons just named, the free elimination being secured by the copious secretion of urine.   Bartels reports
a single case in which uræmic convulsions were present.   In
forty-eight cases reported by Dickinson in which there were
autopsies, there were three instances of uræmic convulsions,
two of coma, and one of unnatural drowsiness.

But lardaceous disease of the kidney never occurs alone.
It is always accompanied by similar changes in the liver,
spleen, and often of the intestinal canal.   Hence, evidences of

alterations in these organs are more or less marked. Thus the percussion areas of the liver and spleen are almost always enlarged, and the bloodvessels of the stomach and intestines are often involved. In the former event obstinate *vomiting*, and in the latter equally obstinate *diarrhœa* results. The latter is far more frequent than the former.

As to *duration*, the disease generally runs a very chronic course, which is limited only by the disease of which it is a complication. As such it is always of shorter duration than interstitial nephritis, and may be shorter than chronic parenchymatous nephritis, although the latter affection and lardaceous disease more closely resemble each other in respect to duration. It is only reasonable, that when superadded to previously existing exhausting disease, the two would hurry an issue more rapidly than either alone. Yet the renal affection is subject to the same improvement to which the general or local one may be subject. I have now under my observation a case of chronic phthisis in which lardaceous disease has been recognized as present for over two years, while the phthisical one has existed much longer, and the patient's condition is in all respects better than it was two years ago. When obstinate diarrhœa and vomiting supervene the end usually is not remote.

Complications.

So much has already been said of what may be properly called the causal complications of lardaceous disease that they require no further mention than enumeration in this connection. They are phthisis, syphilis, caries, necrosis, long-continued suppuration from any cause, and long-continued exhaustive drains upon the system, including albuminuria.

As common results of the same causes, rather than as complications, are to be mentioned amyloid disease, with enlargement of the liver and spleen, and amyloid disease of the bloodvessels of the gastro-intestinal tract, with the vomiting and diarrhœa resulting from these. After these, lardaceous disease is subject to the complications of the other forms of renal disease, but they are less numerous. Bronchitis occurs most frequently. Pleural, pericardial, and peritoneal effusions are very rare, as are also hypertrophy of the heart and retinal changes. Endocarditis, erysipelas, and epistaxis have occurred.

Diagnosis.

There are some instances in which lardaceous disease is easily recognized. If a patient has had syphilis with secondary and tertiary symptoms, or has long been a victim to phthisis, and he is discovered to be œdematous, and to have a large albuminuria, with waxy hyaline and fatty casts, and an enlarged liver and spleen, and obstinate diarrhœa, there can be little doubt but that lardaceous disease is present. But where neither of these two general diseases are present, or the phthisis has not existed a very long while, or there is not decided evidence of enlarged liver and spleen, we cannot be certain. While it is never safe to diagnose lardaceous disease without the presence of enlarged liver and spleen, the presence of these enlarged organs along with large albuminuria, and the other symptoms which attend it, do not necessarily imply lardaceous disease. The symptoms and course of the disease, particularly in its latter stages, are so like those of chronic parenchymatous nephritis that it is sometimes impossible to distinguish the two. Further, there is every reason to believe that chronic nephritis is sometimes caused by the same dyscrasic conditions as produce the lardaceous disease. In such cases, therefore, a diagnosis is impossible. Dr. Paul Fürbringer[*] has recently, in an article already alluded to, reported four cases which so admirably illustrate these difficulties that I feel justified in occupying the space necessary to mention them. In all four cases there was chronic phthisis with ulceration of the bowels. In all four he was fortunate enough to be able to watch the appearance of that series of symptoms which are accepted as pointing to the diagnosis of amyloid disease of the kidney. In

* Zur Diagnose der amyloiden Entartung der Nieren, Virchow's Archiv, Bd. 71, 1877, p. 400.

all four was there splenic and hepatic enlargement, and in all four, autopsies were made. In case one, neither the kidney, liver, nor spleen responded to the iodine test, but the kidney and spleen responded to the anilin-blue test. There was, therefore, amyloid disease, which was uncomplicated by chronic nephritis. In cases two, three, and four, there was no reaction on the part of kidney, liver, or spleen to either test. In case two, there was neither amyloid disease nor chronic parenchymatous nephritis,* and cases three and four were well-marked cases of parenchymatous nephritis.

Finally, the two conditions may exist jointly, where a parenchymatous nephritis may be ingrafted on a lardaceous kidney; or the two may result from the same cause, or possibly the lardaceous disease may be the result of the exhaustive albuminuria of a previous chronic nephritis, as originally suggested by Dr. George Johnson. So that in a certain number of cases, also, the separation of the two is impossible.

When these combinations do not exist the urine in lardaceous disease is more abundant and lighter in color, casts are fewer and hyaline, with a few oil-casts, while in parenchymatous nephritis they are more numerous and include every variety. Blood-corpuscles are rarely found in the urine of lardaceous disease, while they are occasionally present in that of chronic nephritis. If the original disease, the possible cause of a lardaceous disease, has entirely disappeared before the renal malady was detected, the chances of the latter being lardaceous disease are decidedly fewer.

The only other form of renal disease which it is at all possible to confound with lardaceous disease is interstitial nephritis. But in this we have the almost total absence of dropsy, small albuminuria, and scanty sediment, in which granular and hyaline casts are found. While the quantity of urine is increased in both these forms of chronic Bright's disease, the quantity is larger in interstitial nephritis. Hypertrophy of the left ventricle, an almost invariable symptom in contracted

---

* Fürbringer does not say what form of kidney disease existed here, but leaves us to suppose that the kidneys were normal.

kidney, is very rare in lardaceous disease, while enlargement of the spleen and liver are common, and do not occur in interstitial nephritis.

Contracted kidney may also be associated with lardaceous disease. With regard to this relation, Bartels says he would prefer to consider the gradual cachexia produced by the genuine contracting kidney to be the primary mischief and the cause of the secondary amyloid degeneration of the vessels of the kidney. Why he should admit this, and yet not admit that the same result may be brought about by the cachexia of parenchymatous nephritis, which is so much more marked, I cannot understand. I should prefer to consider these two either an accidental complication, or the interstitial disease secondary to the lardaceous disease, the amyloid material acting as an irritant and producing the interstitial hypernucleation and fibrosis in the manner already alluded to.

Prognosis.

In the matter of prognosis much depends upon the presence or absence of the original disease causing the lardaceous change in the kidneys. If the former cannot be cured the effect of the latter can only be to hurry on the unfavorable termination of the former, although it is subject to the abatements as well as exacerbations of that affection. If the original disease is curable, and the patient young, there are no limits to the possible improvement, although it is scarcely likely that the diseased structures are ever restored to their normal state. But as it is unlikely that all the renal vessels are involved in the change, and the organ itself, especially before its complete development is attained, is one capable of assuming an extraordinary degree of supplemental function, it is not impossible that there may be a complete restoration.

If the patient is past middle life, even if it should happen that the original disease has disappeared, the probabilities of recovery are a minimum, while a decided degree of improvement is not impossible. If the stage of alteration of the bloodvessels of the stomach and intestines, as attested by obstinate vomiting and diarrhœa, is reached, the disease is necessarily rapidly fatal.

Treatment.

Of lardaceous disease it may be said with greater emphasis than any other form of renal disease, "an ounce of prevention is worth a pound of cure." A due appreciation by surgeons and syphilographers of the causes of lardaceous disease would prevent the occurrence of many cases; the timely amputation of a limb, long the seat of suppuration, and the thorough treatment of syphilis being all that is necessary to accomplish this. To this end also frequent examinations of urine should be made by the surgeon in charge of cases of the kind so often referred to, and the slightest indication of albuminuria should be the signal for prompt interference, if such be possible, while the possibility of the occurrence of this renal complication should always be before the surgeon's mind.

In syphilis the faithful and persistent use of remedies for a sufficient time after all symptoms of the primary and secondary affections have disappeared is essential. From a somewhat careful examination of the subject, rather than from personal experience, I am satisfied that the "continuous," rather than the "intermittent," treatment of syphilis, by small doses of mercurials long continued, is the plan most likely to secure the eradication of the disease, and subscribe heartily to the dictum of Dr. E. L. Keyes, announced in his paper read before the International Medical Congress* in Philadelphia, 1876:

"I think that a case treated from the first symptom should receive mercury continuously *in small doses* for a period not less than two and a half years, or, in any event, until at least six months have passed after the entire disappearance of the clearly syphilitic symptoms." By small doses are meant doses of $\frac{1}{35}$th to $\frac{1}{50}$th of a grain. This is practically the treatment of Ricord, although he would also use iodide of potassium six months after he had discontinued the use of mercury. The iodide being a remedy rather for the symptoms than a cure for the disease, should be administered in sufficient doses

---

* E. L. Keyes, Treatment of Syphilis, Transactions of the International Medical Congress, Philadelphia, 1877.

as required, in fact interruptedly, while the mercurial treatment should be continued for the period named.

With regard to phthisis, a greater stimulus to our efforts both to prevent and cure than already exists, is scarcely possible, while the futility of those efforts in the very large number of cases is very well known.

If the causing disease continues to exist, the treatment of the lardaceous disease is the treatment of the former,—if it is syphilis, iodide of potassium and mercurials; if phthisis, cod-liver oil, iron, quinia, an abundance of nourishing food, in which milk and cream should be conspicuous, alcohol and restorative measures generally, together with fresh air and suitable exercise.

Supposing the original disease to have disappeared, the measures of treatment indicated are precisely those of parenchymatous nephritis, for the details of which the reader is referred to the chapter on that disease.

As to special modes of treatment directed to the disease, I have little confidence in them. Bartels recommends the use of iodide of potassium as curative of the disease, believing he is indebted for the favorable issue of the cases which have recovered under his care to that remedy, although he does not know how it acts upon the diseased walls of the vessels. He supports, at the same time, the general nutrition by preparations of iron, vigorous diet (meat and milk), and good wine, and maintains sufficient action of the skin by baths. Dickinson also recommends iodide of potassium when the disease is of syphilitic origin, directing the remedy to the affection causing the lardaceous condition rather than to the latter. He also endeavors to compensate the loss which the system suffers in exhaustive suppuration, not by general diet alone, but by salines, chiefly potash salts, and has reason to be well satisfied with the results obtained. Liebig's extract of beef should be especially useful under these circumstances, as it supplies the saline constituents of beef in a concentrated form.

The treatment of complications is in no way peculiar, and has been so fully covered in previous sections that no further allusion is here required.

SECTION VIII.

INTERSTITIAL NEPHRITIS.

Synonyms.—Contracted kidney, chronically contracted kidney, renal cirrhosis, cirrhotic kidney, granular degeneration, granular kidney, red granular kidney, gouty kidney, renal sclerosis.

INTERSTITIAL NEPHRITIS IS A SLOW INFLAMMATORY PROCESS SEATED IN THE INTERSTITIAL TISSUE, WHICH, BECOMING ABNORMALLY INCREASED, SUBSEQUENTLY CONTRACTS AND DESTROYS THE TUBULAR ELEMENTS OF THE KIDNEY; REDUCING ITS SIZE AND PRODUCING OTHER CONDITIONS MORE OR LESS COVERED BY THE SYNONYMS, OR MORE CLOSELY DESCRIBED IN THE SECTION ON THE MORBID ANATOMY OF THE DISEASE.

Etiology.

Of all the recognized forms of Bright's disease, interstitial nephritis presents the largest number of instances in which the cause is undiscoverable. There are, however, some well-determined causes.

Among the most tangible of these is *gout.* Gout is associated with so many cases of contracted kidney, that the term *gouty* kidney has come to be a well-recognized synonym for the product of interstitial nephritis. There are probably no cases of gout, which have continued for any length of time, which are not accompanied by this condition of the kidney. It would seem to be the accumulation in the blood of the poison of gout,—the excess of uric acid and its compounds,—which causes the interstitial inflammation.

Another well-recognized cause is *lead poisoning.* The lead in the blood, like the essential poison of gout, being the irritant which excites the proliferating activity of the connective tissue elements. Hence painters, glaziers, workers in lead in any shape, are frequent victims. Dr. Dickinson considers it safe to assert that, of painters, at least one-half eventually die of granular degeneration of the kidneys.

Long-continued *cystitis*, especially following gonorrhœa, is undoubtedly a cause in a few instances, the inflammation travelling up the ureter to the pelvis of the kidney and thence to the intertubular tissue.

Among the causes, the operation of which cannot be so directly proven, is *mental anxiety*, whether the result of grief or of business and financial cares. Dr. Clifford, quoted by Dr. Robert T. Edes,* goes so far as to attribute " twenty-four out of thirty-two cases, in private practice, to some long-continued anxiety or great grief." Certain it is that this disease very often exists for a long time undiscovered in business men who have lived under a state of constant mental tension.

Hereditary influence is occasionally 'a cause of the contracted kidney. A remarkable instance of this hereditary tendency has recently come to my notice. I was recently consulted by a gentleman, aged 30, who has granular kidneys. His father and mother both died of Bright's disease, aged 56 and 63 years respectively. The mother had convulsions. A brother died of Bright's disease, without convulsions, at the age of 37. Two children of this brother had Bright's disease when 4 and 7 years of age. A second brother died at the age of 29 with convulsions. A third and fourth brother, aged 23 and 32 years respectively, have had Bright's disease for six years. A sister, aged 36, has had Bright's disease for five years. A brother, aged 26, and a sister, aged 34, have as yet no signs of Bright's disease. A maternal cousin died of undoubted Bright's disease, and other members of the family, belonging to previous generations, died with symptoms which suggest Bright's disease. The patient himself has undoubted granular kidney, discovered in August, 1880. An examination of his urine in 1876 revealed no evidences of the disease. There is no gout in the family. Dr. Dickinson also relates the history of a

---

* Edes, Robert T., Some of the Symptoms of Bright's Disease. Boston Med. and Surg. Jour., vol. ciii, No. 2, July 8th, 1880.

family in which a hereditary albuminuria existed, apparently independent of gout.

Alcohol, formerly thought to be a potent cause of the cirrhotic kidney, is now acknowledged to be an infrequent one. The analogy of this condition to the cirrhotic liver suggested a similar irritant action of the alcohol in the blood upon the interstitial tissue of the kidney. But although the portal blood contains a large amount of alcohol after its liberal ingestion into the stomach, by the time the blood passes through the heart and lungs and gets into the kidney, very little if any remains unoxidized. It is barely possible, however, that when enormous quantities are used, enough may remain in the blood passing through the kidney to irritate its connective tissue, and also the cells lining the tubules. The latter, from their efforts to remove it, are probably more frequently irritated than the former; whence also the possibility, though rare also, of chronic parenchymatous nephritis being caused by it, as referred to in discussing the latter disease.

Valvular disease of the heart is another cause of interstitial nephritis, which is the immediate result of the venous congestion shared by the kidney in conjunction with all abdominal organs, when there is insufficiency of the heart's valves. But while the kidney is always altered in every case of valvular disease which has existed for any length of time, yet it is seldom granular. The kidney most frequently found at autopsies of cases of valvular disease is hard, dark-red or bluish, congested and somewhat larger than the normal organ, but smooth, not granular. This is called the cyanotic kidney, the kidney of cyanotic induration, or of passive congestion. On section, the same vascular appearance is seen, the pyramids being more deeply colored than the cortex. Minute examination reveals the capillaries and veins distended with blood, as far back as the Malpighian bodies. The epithelium of the tubules is unchanged or perhaps slightly more granular than in health. The *interstitial tissue is but slightly increased, but tougher and more distinct than in health.* Now this is the condition of the kidneys found in most cases of valvular heart

disease which have continued, say a year or more.   But if the
disease be still more prolonged we have the essential pathological condition of the granular kidney.   There is a positive
overgrowth of the interstitial tissue, which is especially evident about the Malpighian bodies.   Contraction takes place,
interfering with the nutrition of the epithelium of the convoluted tubules, which become granular, fatty, disintegrated,
and portions of the tubes themselves and their Malpighian
bodies are destroyed, producing cicatricial depressions, and
an uneven granular surface.   It is true that the granula-
tions are never so large or distinct as in the typical granular
kidney, nor is the organ ever so much reduced in size, perhaps never below the normal size.   But the mechanism of
the production of the condition is precisely that of the ordinary granular kidney.   It is simply a difference in degree.
Therefore I do not hesitate to place valvular disease of the
heart in the category of causes of interstitial nephritis.   In
this view I am sustained by Dickinson and Lecorche.   The
former,* in 153 post-mortem examinations of persons dying
with valvular disease of the heart, found the kidneys in 67
having granular surfaces, and more or less contracted cortices.
In 29 the kidneys were hard, congested, and increased in bulk,
but still smooth.   Dr. Barclay,† in an analysis of 79 cases of
valvular disease, found 28 having granular kidneys.   Bartels,
on the other hand, prefers to consider it a separate process,
chiefly, if I understand him aright, on the ground of its slow-
ness or chronicity.   He compares it with brown induration of
the lung as it occurs in connection with valvular heart disease,
or the thickening of the peritoneum which follows upon portal
congestion.   I do not think simple duration sufficient to separate one process from another with which it is otherwise identical.   Of course this condition of the kidney as the result of
heart disease is to be distinguished from that in which the renal
disease is primary, as occasionally happens.

* Dickinson, op. citat., p. 155.
† Med.-Chirurg. Transac., vol. xxxi, p. 96.

Pregnancy is rarely, if ever, a cause of interstitial nephritis. I have already named it among the fertile causes of acute and chronic parenchymatous nephritis, being here in accord with Bartels and against Dickinson. Indeed I am unable to understand how the latter writer draws the conclusions which he does from the cases he reports. For the morbid appearances he describes seem to me more those of acute and chronic parenchymatous nephritis than of interstitial nephritis, some of those which most resembled the granular kidney being in the stage of contraction.*

Age and Sex of those Affected.—Interstitial nephritis is commonly considered a disease of middle age. And so it is in the sense that the majority of persons in whom it is discovered are past forty. And I confess to some surprise when, on investigating the matter, I found that there were so many cases, although still a small number under thirty. The youngest patient whom I have had was twenty-six. But it will be seen from the appended tables by two of the largest observers, Dickinson and Bartels,† that the number of cases occurring under forty years of age is not inconsiderable. The figures indicate the age at which death occurred, and there was an autopsy in each case.

BARTELS.	DICKINSON.
1 was 18 years.	1 was between 11 and 20 years.
1 was 19 years.	24 were between 21 and 30 years.
2 were 20 years.	50 were between 31 and 40 years.
4 were between 20 and 30 years.	93 were between 41 and 50 years.
9 were between 30 and 40 years.	76 were between 51 and 60 years.
4 were between 40 and 50 years.	47 were between 61 and 70 years.
7 were between 50 and 60 years.	17 were over 70 years.
5 were between 60 and 73 years.	
33	308

* The reader who is sufficiently interested in the subject is referred to a paper by the writer on "The Causal Lesions of Puerperal Convulsions," read before the Pathological Society of Philadelphia, April 28th, 1878, and published in the Transactions for 1878-1879. The entire subject of Puerperal Convulsions in its relation to albuminuria and Bright's disease is here discussed.

† Bartels, op. citat., p. 411. Dickinson, op. citat., p. 145.

It must be remembered that there is a tendency to overgrowth in the interstitial tissue of the kidney as of other organs in old age. Hence the term *senile atrophy* of the kidney. It is not safe, therefore, to call every instance of atrophied kidney met in the post-mortem room a case of interstitial nephritis. The clinical history, or some one of the well-marked symptoms of the disease, as albuminuria or uræmic symptoms should have preceded to sustain the diagnosis. This was the defect in the method followed by Gull and Sutton,* who, in the post-mortem examination of a large number of cases, independent of previous disease, noted that out of thirteen cases examined between the age of sixty and seventy, twelve had contracted kidney. At the same time, this atrophy of old age may proceed to such degree as to attain positive pathological degrees, manifested not only by the morbid anatomy of the kidney, but by the symptoms of interstitial nephritis.

As to *sex*, nearly twice as many males are subjects of the disease, because of the more frequent exposure of the former to the causes of the affection.

Morbid Anatomy.

In interstitial nephritis, both kidneys are involved, but there is often a marked difference in the amount of disease in each.

Macroscopically the organs are evidently smaller than in health, often less than half as large. I have occasionally seen them less than five centimeters (two inches) in length. Next to this reduction in size, the most striking feature of the contracted kidney is its uneven or granular surface, which is, however, not always recognizable until after the capsule is removed. Very characteristic also is the presence of *cysts* with more or less clear watery or gelatinous contents, often visible through the capsule. These are not invariably, but quite constantly present. The *capsule*, itself thickened, strips off with difficulty,

---

* On the Pathology of the morbid state commonly called Bright's disease with contracted kidney (Arterio-capillary fibrosis). Med.-Chirurg. Transac., 2d series, vol. xxxvii. 1872.

dragging portions of the secreting structure with it. Owing to the resistance which the blood meets in its passage through the kidney, as the result of pathological alterations to be presently described, a larger portion of it passes out of the organ by way of the capsule. Hence the bloodvessels of the latter are dilated, as are also the lymph spaces.

Bereft of its capsule, the kidney is granular, hard, tough, and usually darkened in color; whence one of its names, the "red granular kidney." This color is in strong contrast to the white or slightly yellow tinge of the *fatty and contracting* kidney, and although not always marked, but sometimes even substituted by a paleness, it is still easily distinguished from that of the contracting kidney of parenchymatous nephritis.

The *granules* on the surface of the contracted kidney are distinct round and oval elevations of the surface, ranging in size from that of a pin's head to a pea, or from one to five millimeters ($\frac{1}{25}$ to $\frac{1}{5}$ of an inch). Those of smaller size are most numerous, and at first correspond with the lobules, the bases of which are visible on the surface of the normal organ. The larger ones result from the coalescence of two or more smaller. The granules themselves are of a lighter color than the depressed circlets between them, which are tinted with vascularity, and have a purplish or faint red hue.

The *cysts* already referred to are now more distinct (after removal of the capsule), and vary also greatly in size. While equalling in minuteness the smallest of the granules, some of them are as large as a walnut. The larger are apt to be ruptured on stripping off the capsule.

On section it is at once evident that the reduction in size of the kidney is almost wholly due to a narrowing of the cortex, although the medulla is also contracted. The former may not be more than three or four millimeters ($\frac{1}{8}$ to $\frac{1}{6}$ inch) in width, and exhibit every degree between this and the normal. The Malpighian bodies are smaller, less numerous, and can scarcely be detected by the naked eye, while the small arteries are more prominent from thickening of their walls. The increased density and firmness of the organ are apparent. If the conditions occur in a gouty subject, linear chalk-marks of urate of soda may be present, more particularly in the pyramids of straight tubules, and are contained within as well as between the tubules. The little cysts in varying numbers are scattered throughout the section from cortex to papillæ, but they are more numerous in the former. Rindfleisch says the cysts in the medulla, in the neighborhood of the papillæ, are formed in the flexures of the looped tubes, where they originate in the colloid metamorphosis of cylinders of fibrin. The cysts may be altogether wanting.

The *pelvis of the kidney* may be unaltered. It is sometimes enlarged from retraction of the pyramids. The calyces may be elongated. On the other hand if the kidney is very much reduced in size, the capsule may be pursed up, and proportionately smaller.

Da Costa and Longstreth, in a recent article,* call attention to the constant presence of an increased amount of fat surrounding the contracted kidney, even in emaciated subjects.

Minute Structure.—Minute examination of thin sections through the cortex, clearly reveals the change to be an overgrowth of connective tissue, with destruction of the tubules and bloodvessels. The process is best studied if the sections include the capsular edge, as it progresses from without inwards. In these may be seen extensive tracts of connective tissue of different ages, separating the tubules, which, in sections of healthy kidneys, are seen to be closely in contact without appreciable intertubular substance. The younger connective tissue exhibits a simple hyperplasia of cells, with a scanty transparent intercellular substance, while that which is older contains fewer corpuscles and a more or less distinctly fibrillated intercellular substance.

The *tubules* themselves, appear, in places, quite normal; in others they are represented by fragmentary portions in which the cells are still unchanged; in others again the cells exhibit a granular degeneration; some are evidently dilated;

---

* Researches on the State of the Ganglionic Centres in Bright's Disease, Am. Jour. Med. Sci., N. S., vol. 80, 1880, p. 25.

others still are completely shrivelled, while it is evident from the larger areas of connective tissue that many have completely disappeared. In a few tubules, waxy casts are present. The Malpighian bodies are surrounded by concentric layers of nucleated connective tissue. Many of them are shrivelled and atrophied, and an attempt to inject them with colored injecting fluids fails either partially or completely. Some thus altered lie detached from the tubules with which they should be continuous.

The *cysts* originate partly in dilatations of obstructed segments of the uriniferous tubules, and are partly the result of a dilatation of the Malpighian capsules. Proof of the latter mode of origin is found in the fact that compressed capillary tufts are sometimes found lying up against one side of the wall of the cyst.

The same overgrowth of connective tissue takes place in the *medulla*, but it appears here later, extends more slowly, and never reaches the degree found in the cortex. For this reason the contraction is not nearly so great, but the medullary area is still somewhat diminished by retraction of the pyramids. The excreting tubes in the medulla are in some places dilated, and in others twisted.

Minute examination of the granules on the surface of the kidneys shows them to be made up of tubules in a tolerably perfect state, the mass having been caused to project by the constriction of the connective tissue encircling its base.

The *vascular system* of the contracted kidney is the seat of important changes. In the first place it shares with the tubules the compressing effect of the contracting new formation. As the result of this, a part of the capillary system is destroyed, and in the part thus destroyed are many capillary coils in the Malpighian bodies. Hence, as many afferent arterioles send their blood directly into the second capillary network, which is also cut down by the pressure. The vessels which remain are often dilated and twisted, and in consequence of the destruction of numerous Malpighian bodies, send much of their blood out through the capsule.

A further very striking vascular alteration is a hypertrophy, with subsequent degeneration of the muscular coat of the arteries and arterioles, and thickening of the external fibrous coat. This alteration has given rise to a good deal of discussion. Dr. Bright noted changes in the renal artery in kidney disease, but Dr. George Johnson, in 1867, described the change more thoroughly, and showed it was common to the whole body. He considered it a pure muscular hypertrophy,—a reflex reactive overgrowth, whose object is to resist the entrance of the contaminated blood into the tissues. Drs. Gull and Sutton, in 1872, showed that along with the muscular hypertrophy there was also a thickening of the external fibrous coat. They, however, made the condition of the renal arteries a part of a general one, affecting the entire arterial system, which, while it commonly begins in the kidney, may begin primarily in other organs. Indeed, according to them, the kidneys may be but little if at all affected, whilst the morbid change is far advanced in other organs. They characterize the condition as a "hyaline fibroid" formation in the arterioles and capillaries, and suggest for it the name "arterio-capillary fibrosis." They say further, that the morbid state is allied to the conditions of old age, and that we cannot refer the vascular changes to an antecedent change in the blood due to defective renal excretion, but that they are due to distinct causes not yet ascertained. The element of thickening of the external coat, in addition to the hypertrophy of the muscular, these observers have established; but their view as to its being but a part of a general condition, and in no way the result of kidney disease, is not accepted, so far as I know, by any. Nor have their views as to its causation been any more favorably received.

This condition is invariably associated with *hypertrophy of the left ventricle* of the heart. The two increase *pari passu*, and it is therefore impossible to disassociate their consideration. That they are both the *result* of the renal condition it seems to me requires no further proof than that furnished by Johnson and Dickinson, in their respective treatises and papers.

As to the immediate cause of this hypertrophy and overgrowth, Dr. Johnson's view best explains the facts, while it is consistent with the modern physiology of the arteries. It has already been said that he ascribes the thickening of the middle coat to a reactive overgrowth. This is stimulated to occur in two ways. The arterioles contract forcibly to resist the admission of the blood poisoned with the accumulation of excreta usually eliminated by the kidneys. This is the first stimulus. The heart then contracts with additional power to overcome this resistance, and thus it hypertrophies. But the arterial vessels contract still more forcibly to resist this impulse, whence the second stimulus to overgrowth of the muscular coat. As Johnson considers the change purely a muscular one, he is not called upon to explain the thickening of the adventitia. But the same causes which excite the hypertrophy are capable of exciting an inflammatory overgrowth of the connective tissue which forms the outer sheath, whether it be increased blood pressure or the irritant character of the blood. Both probably co-operate to produce it. So far as the arteries in the kidney itself are concerned, it is very likely that the thickening of the external coat is partly due to the inflammatory changes which, in this form of disease, select the connective tissue of which it forms a part.

In the last edition of his admirable treatise on the *Pathology and Treatment of Albuminuria*,* Dr. Dickinson, in advocating the older view of Bright, that both the cardiac hypertrophy and thickening of the muscular coat of the arteries are the result of a *capillary* resistance, points out what he believes to be a serious objection to the view of Johnson. The office of the muscular coat, says Dickinson, is not to resist, but to aid the heart in the propulsion of blood; the force of propulsion, which begins with the cardiac systole, being continued as a wave of contraction, which passes in a peristaltic manner from the orifice of the aorta throughout the arterial system. It is true, this may be one of the offices of the muscular coat of arteries, but it is secondary to a much more important and better determined one of regulating the calibre of the bloodvessel, in accordance

---

* Page 328.

with the temporary demands of an organ or tissue for blood. This is a reflex operation, controlled by the vasomotor nerves, and although it resists the heart in one sense, this resistance is never sufficient to induce hypertrophy of the left ventricle, because, while the blood may be partially cut off from an organ or set of organs, by the stopcock action of the muscular coat, all the remaining arteries of the system are relaxed and ready to receive the extra blood thus shut off. But in the case of the poisoned blood, from insufficiency of secretion by the kidneys, *all the arterioles* are resisting its admission. There is no free outlet. The heart, in its efforts to overcome this resistance, hypertrophies in its left ventricle, against which the arteries again react, and thus the antagonism keeps up, and an increased hypertrophy results.

Finally, all physiological studies go to prove that the capillaries do not possess an active power of changing their calibre and, therefore, of resisting the admission of blood. And although the destructive compression of capillaries, and probably even arterioles which occurs in the kidney itself, in cirrhosis of that organ, obstructs the movement of blood through it, such local resistance is not sufficient in the presence of the unobstructed state of the remainder of the arterial system to produce a *universal* hypertrophy of the muscular coat including that of the heart.

The degeneration alluded to, as succeeding the thickening of the bloodvessel wall, are of the kind which usually succeed hyperplasias, viz., fatty. In the muscular coat the nuclei of the fibre-cells waste, and finally disappear, leaving the coat recognizable only by its transverse striation, although this appearance may also be obscured by the presence of fat-drops, and even crystalline matter. The subsequent changes in the external sheath consist in the conversion of the thickened fibroid adventitia into a more hyaline material.

The final effect of these alterations is to produce a brittleness in the arteriole walls, which disposes them to rupture on very slight increase of intravascular pressure. Hence, the frequent fatal termination of cases of interstitial nephritis by apoplexy.

Drs. Da Costa and Longstreth, in a recent paper,* have described as more or less constant in Bright's disease, and especially in the contracting kidney, certain changes in the nervous renal ganglia, which consist essentially in a hyperplasia of the connective tissue and a fatty degeneration of the nerve-cells.

They say that while this lesion might be looked upon as forming a part of the general process of degeneration in connection with the kidney disease, they think it is the *cause* of the renal malady and precedes the degenerative changes.

Also that the diseased condition of the ganglia furnishes the clue to the alterations of the vessels of the kidneys.

And finally, that similar changes producing similar results may exist in other ganglia ; for instance, in the cardiac plexus, explaining the hypertrophy of the heart.

Da Costa and Longstreth in the paper alluded to, say they do not think the heart hypertrophies because of the opposition the passage of the blood meets in the renal circulation, but that it is to be traced to a central origin ; in one case to the cardiac ganglia, and in the other to the renal. "The same statement applies to the changes in the vessels and tissues in other organs of the body, and when they in their turn are affected the ganglia which preside over them as centres most probably have been acted upon in a similar manner as the ganglia already mentioned."

For a description of the *retinal changes* which occur most frequently in this form of Bright's disease, see the following Section IX, by Dr. William F. Norris.

Symptoms and Course of Interstitial Nephritis.

The degree of uncertainty as to the origin of a large majority of cases of contracted kidney is only equalled by that of the insidiousness of their approach. The beginning of the disease is certainly not marked by any distinctive symptoms; and its progress is often unmarked by any, until those of uræmia, as manifested by the unexpected convulsion, drowsiness, or stu-

---

* Amer. Jour. of the Med. Sciences, N. S., vol. 79, July, 1880, p. 17.

por, mark the beginning of the end. The attention of the
physician being thus directed to the urine, albuminuria and
casts are discovered and confirm the condition. To the observ-
ing physician, perhaps an unaccountable weakness or slow
convalescence from acute disease, or other unexplained condition, may suggest an examination of the urine; or the peculiar tense and bounding pulse of hypertrophy of the left ventricle, or the more tangible symptom of a slight swelling of the
feet or ankles, recognizable only at night or through the unexpected tightness of a boot, may lead to the same examination.

Attention being called to the *urine*, it will be found to present characters which are more or less distinctive, and commonly in connection with the absence of other symptoms, alluded to as so characteristic, lead easily to a diagnosis. It is,
when freshly passed, acid in reaction, copious, often exceeding
the normal amount, never scanty except in the last stages of the
disease. The quantity may reach 2700 cc. (90 oz.), but this is
very rarely. The patient very commonly has to rise at night,
probably not more than once or twice, to pass his water. Consequently, the urine is light in color and of low specific gravity,
—1010–1015,—and contains a trifling or moderate flocculent
sediment. It is *generally albuminous*, but the albumen is
small in amount, and may be temporarily absent, or it may be
absent before a meal, and present after it. Later, however, it
becomes constant. It seldom exceeds one-fourth the bulk of
fluid tested, and is very constantly a good deal less, showing
a delicate line of white by Heller's nitric acid test.

Tube-casts are present, but not usually numerous. They are
almost solely hyaline, and pale granular. Some of the hyaline
casts are delicately so, requiring nice illumination for their detection; others are distinct and sharply cut; others, still, contain two or three glistening oil-drops. Casts may at times be
absent and again reappear, as is the case with albumen. Towards
the termination of cases of interstitial nephritis the urine diminishes in quantity, the specific gravity increases, and the casts
become much more numerous, and include among them highly
granular or dark granular, and occasionally even blood-casts,

in addition to those alluded to. In a case which recently terminated under my observation, the specific gravity rose to 1030, and the urine diminished to about 25 cc. (8 oz.). The low specific gravity of the urine in the early stages is largely the result of the increased quantity, which often equals 1800 to 2100 cc. (60 to 70 oz.), and even rarely 2700 cc. (90 oz.).

The *urea* is also diminished sooner or later, and in this manner the lower specific gravity is contributed to. This diminution becomes marked towards the close, accounting for the uræmic symptoms which often first announce the disease. It may be as low as 1 gram (15 grs.), and may be anything between this and the normal 24 hours' quantity, which may be put down at from 20 to 40 grams in an adult.

All the remaining normal constituents may be said in general terms to be diminished.

As to the other symptoms which were alluded to as suggesting an examination of the urine, a feeling of unaccountable *weakness* or *of being tired* was mentioned. This is very often present, but it is a symptom which is present in many conditions, and should only be considered as suggestive.

Slight œdema about the feet and ankles is often present, being so slight as to escape detection, or is discovered accidentally. When present it is significant, but it is often entirely wanting. We have the explanation of this slight œdema or its entire absence, in the free secretion of water, which is such a decided symptom of the early stages of this disease.

Hypertrophy of the left ventricle of the heart is so constant that it may be properly considered as a symptom rather than a complication, which it strictly is. No case of interstitial nephritis has existed for any length of time without the presence of this condition, and as few cases are discovered until they have existed a good while, few are found without hypertrophy, although it is not always marked. In about one-half the cases, however, there is evident hypertrophy. Corresponding to this, the pulse has a peculiar tense and bounding character. These two symptoms have, therefore, a diagnostic value. I have already alluded to the cause of this hypertrophy,—the in-

creased effort of the heart to propel the blood through capillaries which resist its presence. Ordinarily there are no
symptoms of this enlargement except the increased percussion
area and the peculiarity of the pulse. Occasionally, however,
there are symptoms of cardiac distress, dyspnœa, and palpitation. Also accentuation of both sounds and reduplication of
the first. The latter is probably due to a want of synchronism
in the systole of the two ventricles.* There is usually no
murmur because there is no valvular disease. The latter may
be present, however, in one of two ways. The patient may
have had valvular disease prior to the renal malady, or the
latter itself, by its long continuance, have produced endocarditis and atheroma which are among its results, particularly in
those who have passed middle life, thus increasing the natural
tendency of age to this condition.

Increased arterial blood-pressure is another symptom, which
is the direct result of the hypertrophy of the left ventricle.
This continues as long as the power of the heart lasts. When
the latter begins to fail, the blood-pressure diminishes, and
with it begin a train of symptoms, among which diminished
secretion of urine and dropsy are the most conspicuous. It occasionally happens that, although this hypertrophy is a compensating measure, it is the cause of inconvenience, manifested
by the cardiac distress, dyspnœa, palpitation, and dizziness,
alluded to as sometimes present.

The atheroma and the increased arterial blood-pressure are
the causes of another symptom which frequently determines
the mode of death,—rupture of a bloodvessel in the brain; in
a word, *apoplexy*. As stated, this accident occurs generally
late in life, but Dickinson reports a case in which cerebral
hæmorrhage occurred in a girl of twelve. The proportion of
cases of recognized interstitial nephritis in which this occurs is
not large, but many cases of apoplexy are directly traceable
by autopsy to unsuspected renal cirrhosis. Dickinson believes
that of fatal cases of apoplexy one-half are preceded by this
form of disease.

* Sibson, British Medical Journal, April 1st, 1871, p. 238.

Hæmorrhages in other situations are referable to this same altered state of the bloodvessels, as into the retina, from the nose, and even into the stomach. Hence *sudden blindness*, as well as dimness of vision due to retinitis albuminurica, is a symptom which occasionally presents itself.

The latter symptom, *dimness of vision due to retinitis albuminurica*, is a grave one. The pathological alteration in the retina which causes it presents itself usually only at an advanced stage of the renal disease, and is but little remediable. The condition itself I deem of such importance as to demand separate consideration by an ophthalmologist. Accordingly, my friend Dr. William F. Norris has kindly furnished an article on the *Retinitis of Bright's Disease*, which will follow this section.

The termination by *uræmia* occurs more frequently in this than any other form of Bright's disease. Bartels says that nearly all the patients he has seen die in the extreme stage of atrophied kidneys, sank under the symptom of chronic uræmia. I can say the same of my own experience as to uræmia, but acute uræmia has been as frequent as chronic. As already stated, it is frequently the first intimation of the existence of any derangement, and manifests itself in any one or more of the forms already described under acute nephritis. Headache, drowsiness, convulsions, stupor, delirium, maniacal excitement, renal asthma, restlessness, nausea, vomiting, any one of these symptoms may usher in the dreadful train which is so apt to be fatal. But coma or a partial coma is here more frequent than convulsions. Drowsiness and severe headache are also frequent beginnings. Dr. E. C. Seguin has recently called attention to *occipital* headache as a symptom of uræmia.* In a case which terminated recently under my observation, in which delirium, coma, and maniacal symptoms were present, but interrupted by active treatment, the *temperature* was very high, ranging from 102° to 105°, and rose to 107° immediately before death. During the stupor the patient was quiet, and there was no stertor.

---

* Archives of Medicine, vol. iv, No. 1. New York, August, 1880.

Dyspeptic symptoms with obstinate vomiting, particularly in the morning before eating, are apt to usher in a chronic uræmia. Diarrhœa is less common, but also sometimes occurs towards the close, when it may be very difficult to control.

The duration of this form of renal disease is indefinite. Always a chronic process, it may last for years undiscovered, and when discovered before it is too far advanced, the knowledge of its presence will suggest measures of precaution and treatment, which may so prolong life that it need only be determined by its natural limit or some other disease. Yet complete recovery from well-established interstitial nephritis disease is probably unknown.

Complications.

These are, in addition to what have already been considered as symptoms, but of which some might with propriety be classed among complications, bronchitis, pericarditis, pleurisy, pneumonia; and, more rarely, endocarditis, peritonitis, intertubular gastritis, and even inflammation and ulceration of the bowels. But all inflammatory complications except bronchitis, pleurisy and pericarditis are less common than in acute nephritis. Bronchitis is said to be the most frequent complication, while pericarditis is the most dangerous, being almost invariably fatal. The former occurs in about 33 per cent. of the cases, and the latter in 25 per cent. Pleurisy and pneumonia are also of tolerably frequent occurrence, Grainger Stewart finding the former in 15 per cent. of his cases and pneumonia in 7 per cent. The latter author found pericarditis also in only 7 per cent., while Dickinson found recent pericarditis in nearly 25 per cent. ; a proportion which would be much increased if old false membranes and adhesions are included. Acute endocarditis and peritonitis occur but very seldom. It is possible that the obstinate vomiting in some cases may be due to intertubular gastritis, the occasional presence of which is asserted by Drs. Fenwick and Wilson Fox, as distinguished from the follicular inflammation found in acute nephritis. Three cases of ulceration of the ileum are mentioned

by Dr. Dickinson, and one by Dr. Bartels. The occasional
obstinate diarrhœa may be due to this cause.

The cause of this tendency to inflammation attending contracted kidneys is probably the irritant state of the blood, poisoned with the retained excretions which the damaged kidney
is unable to remove.

Diagnosis.

The diagnosis of an interstitial nephritis is usually easy,
if by any means an examination of the urine is suggested.
The increased, or at least undiminished quantity, the low specific gravity, small albuminuria, delicate hyaline, and pale
granular casts, in the absence of other symptoms, are sufficiently
distinctive. The conditions which should suggest such an examination are a feeling of constant weariness, slight swelling
of the feet, drowsiness, intense headache, confused intellect,
dyspeptic symptoms, obstinate nausea, delirium, coma, and
convulsions.

Prognosis.

The prognosis is unfavorable as to complete recovery, but
favorable as to prolongation of life if the diagnosis be made
sufficiently early. If it is not made previous to the setting
in of uræmic symptoms, little may be expected. But even at
this stage energetic treatment may still avail to avert the immediate danger and prolong the patient's life. The possible
sudden occurrence of convulsions and coma, and of death
therefrom, should always be remembered and impressed upon
the relatives of the patient.

Treatment.

From what has been said under prognosis, it is evident that
the most hopeful result to be expected from treatment is the
protection of the patient from the consequences of his malady,
rather than the restoration of the kidney to its normal condition. Our power in the former respect depends largely upon

the stage at which the disease is discovered. But let us first suppose that it is detected, as it often is, at a period in which the urine is tolerably abundant, the albuminuria small, the casts are few, and there is no œdema. Here, again, the indications are, *first*, to maintain the integrity of the blood, chiefly by preventing the accumulation of urea and its allied compounds, and by compensating the loss by albuminuria, which is at first not great; and, *second*, to treat as they arise, the accidents and complications which are often so dangerous to the patient.

The first of these is best accomplished by dietetic and hygienic measures, aided by the use of a few remedies.

First, as to *food*, all that was said under chronic parenchymatous nephritis is more than applicable to interstitial nephritis, because the appetite is still good and a suitable selection can be exercised.

As there stated, it is now generally recognized that the urea found in the blood and eliminated by the kidneys in health, has its chief source in the azotized elements of food, and that the more nitrogenous food we consume, the larger is the accumulation of urea in the blood, and the greater the work thrown upon the kidneys. Now while it is not possible nor desirable to exclude all nitrogenous food, it must be of advantage in this disease to reduce it to a moderate amount. This is accomplished by the substitution of all, or perhaps preferably, of a part of animal flesh by milk and cream, while drawing the elements of a mixed food from the vegetable kingdom. The so-called vegetarians have proved conclusively that it is possible to live and maintain good health upon milk and an otherwise exclusively vegetable diet. And while this diet may not be compatible with the highest mental and physical development of which man is capable,[*] which is not at all proven, the resulting life is perfect enough for all its objects, and will doubtless be acceptable to those who prefer to live. On such a system I have known the patient with contracted kidney to maintain apparently perfect health for many years.

---

[*] See Carpenter's Physiology, 7th English ed., 1869, p. 77.

With regard to *beverages*, there is no doubt that the use of strong alcoholic drinks should be avoided, and brandy, whiskey, and strong sherries and ports should be prohibited. The light wines, and especially the red wines, and lighter alcoholic drinks, as lager beer, porter, etc., may be used.

What was said of *clothing*, *fresh air*, and *exercise*, in connection with chronic parenchymatous nephritis, is even more applicable to interstitial nephritis. Warmth of the body, maintained by *woollen garments next the skin* to encourage its action, and the *avoidance of damp and cold* which check it, are peremptory. The wetting of the body by rain, or of the feet alone, has frequently been the exciting cause of a uræmic attack which was fatal. Rubber overshoes should be worn in all damp weather.

In this connection *sea-bathing* requires to be alluded to. It is well known that sea-bathing sometimes induces albuminuria in individuals perfectly healthy, or at least in individuals at other times free from albuminuria. This must of course be due to a temporary congestion of the kidney, the result of an introversion of the blood from the skin, which is kept up by the length of the bath. Still more mischievous, therefore, must be the effect of prolonged sea-bathing upon one whose kidneys are already damaged and incompetent to perform their office. I am confident that one case of latent contracted kidney that recently came under my observation, was hastened to its fatal termination by this cause, the patient in this instance remaining in the water two or three hours at a time. Sea-bathing, therefore, or indeed any form of cold bathing should be interdicted to the patient with contracted kidney. Sea-bathing is especially mentioned because it is considered so healthful, and involves remaining in the water so much longer at a time. On the other hand the warm bath, and especially an occasional Turkish bath is advantageous.

For the same reason there is no doubt that *residence in a warm equable climate* is often of signal service in interstitial nephritis; and cases are reported where the albumen has disappeared and apparent recovery taken place during such resi-

dence, where the previous duration was such as to make recovery at least improbable.

I recall the instance of a lady past middle life who had
granular kidneys, and who had already begun to suffer the
numerous inconveniences of advanced renal disease. She went
abroad, and, during a residence in Southern Germany, was entirely relieved of all unpleasant symptoms. She described
herself as feeling during that period like a young girl. She
had no sooner reached Liverpool, on her return, than all of these
symptoms returned, and she died in a few months after reaching home. I have no precise knowledge of the frequency of
Bright's disease in the southern part of the United States, but
I seldom hear of cases originating there.

Prolonged bodily or mental fatigue should also be avoided
by these cases, as they have been known to be the exciting
cause of uræmia and death. The patient should live a life as
easy and as free from any of these causes which have been
considered, as his circumstances will permit.

As to *drugs*, they are of limited utility. The tonics, iron,
quinine, and strychnia, are useful and necessary to combat the
tendency to anæmia and weakness, which sooner or later the
albuminuria will induce. Iron is perhaps the most important
and indispensable, and may be given in the shape of the acetate
or the well-known Basham's mixture, of which the formula is
given on p. 138. But it is a question whether the sense of weariness so characteristic, is not due, in a measure, to the accumulation in the muscles of effete matter, which should be eliminated
by the kidneys, as the tired feeling so characteristic of diabetes is
probably due to a retention of sugar in the muscles, the oxidation of which relieves the sense of fatigue. In that event
this weary feeling in interstitial nephritis would be more
likely to be removed by encouraging elimination by measures
which promote the action of the skin. I would, however,
resort to no very active treatment of this kind, but would
recommend an occasional warm bath, or especially a Turkish
bath, with thorough friction and precaution from cold ; the combination of a gentle diaphoretic, as the sweet spirit of nitre or

the solution of acetate of ammonium, with tonics and chalybeates, or brisk exercise in the open air, with proper precautions against subsequent exposure while perspiring. These, united with sufficient rest, all tend to supplement the action of the kidneys and avert the consequences of their defective action.

Diuretics are not indicated in this stage, because the secretion of urine is already free and not likely to be much increased by further diuresis. The *bowels* should be kept *regular* by the use of the natural aperient waters, the Hunyadi, Friedrichs-halle, and Rakoczy, or an occasional blue pill, but there is no need of decided purgation. The aperient may be combined with iron, and for this purpose sulphate of magnesium or potassium are useful. An ounce of the former, ten grains of sulphate of iron, and a drachm of aromatic sulphuric acid may be added to a pint of water, and of this a wineglassful taken before breakfast. Of course, later in the disease, when the urine is again scanty and we want to bring to bear all available means of increased elimination, both diuretics and purgatives may be desirable to avert the calamity of uræmia. The same principles are to govern us in using them as have already been laid down under acute nephritis. I will mention, however, digitalis as being a diuretic which is at once usually efficient as such, while it will also serve to sustain a failing heart.

The second indication mentioned, the *treatment of the complications and accidents* incident to the condition, resolves itself into the treatment of the bronchitis, the pericarditis, the pleurisy, pneumonia, endocarditis, gastric and intestinal disorders which have been named as occurring, and especially of the most serious calamity of all, uræmia. The treatment of the complications is that of the same conditions under other circumstances. Prompt restorative and even stimulating measures are, however, here earlier called for than in these same conditions uncomplicated, and when to these are added counter-irritation, the chief indications for the thoracic troubles are fulfilled. *Paracentesis* is a measure which is often of signal service in effusions into the chest, and occasionally of the pericardium.

Opium should be cautiously employed, not only in the gastro-intestinal troubles, but under all circumstances, as it undoubtedly increases the dangers of uræmia. This has been abundantly proven. It need not be discarded altogether, for there is indeed no substitute for it in most bowel affections and conditions of severe pain, but it should be given in smaller doses than usual, and its effects watched. In like manner hypnotic, sedative, and antispasmodic effects, when desired, should be produced by *chloral* and *bromides* if possible.

Dyspeptic symptoms are best treated with pepsin and acids. I have found full doses of the wine of pepsin, f$\frac{3}{ij}$ to f$\frac{3}{ss}$., with 5 to 10 drops of dilute nitromuriatic acid, and $\frac{1}{30}$ to $\frac{1}{12}$ of a grain of sulphate of strychnium, here, as elsewhere, among the most efficient remedies for dyspepsia. Or the powdered pepsin in 8 to 10 grain doses may be dissolved in an aqueous acid solution.

Finally, as to the treatment of *uræmia*, when it occurs, the immediate indication is, of course, increased elimination of a decided character. This is accomplished by sudorifics, purgatives, and diuretics. The degree to which these should be pushed will of course depend upon the urgency of the symptoms. If there be simple drowsiness, headache, or gastric derangement, a brisk purge by $\frac{1}{6}$ of a grain of elaterium, or 1$\frac{1}{2}$ drops of croton oil, or a saline, followed by full doses of infusion of digitalis, with acetate or citrate of potash, may suffice to avert the evil. Or, a drachm of jaborandi-leaves may be infused in six ounces of hot water, and a tablespoonful administered to an adult, and repeated if necessary until a sufficient diaphoresis results. If it be desired to counteract any depressing effect, which can hardly exist with so moderate a dose, a drachm of solution of acetate of ammonium, which is also diuretic, may be added to the mixture. The fluid extract of jaborandi, which I have not found so reliable as the freshly prepared infusion, may be given for the same purpose in 20-drop doses. If there is headache, full doses of bromide of potassium are indicated.

Usually, however, when the symptoms of uræmia appear in this disease, they are urgent, and require more active treatment. Coma is very frequently present, and the administration of remedies by the mouth is difficult or impossible. Then $\frac{1}{3}$ or $\frac{1}{2}$ a grain of *muriate of pilocarpin* may be given hypodermically, in solution. I have given $\frac{2}{3}$ of a grain at a single dose, but the occasion was very urgent. A $\frac{1}{2}$ grain is perhaps the most reliable, as I have known $\frac{1}{3}$ grain to be insufficient to produce the sweating, although it generally suffices. If the remedy acts, profuse sweating, with or without salivation and diuresis, will take place within half an hour, and this is often attended with a prompt return to consciousness. If pilocarpin is not at hand, half of the above-directed infusion may be thrown into the rectum, and the remainder in half an hour, if sweating does not result. But while waiting for this effect, the patient may be dry-cupped over the loins and at the back of the neck, and, if necessary, a couple of the cups at the neck may be cut. At the same time a couple of drops of croton oil, diluted with plain oil or glycerin, may be dropped into the mouth, or $\frac{1}{4}$ of a grain of elaterium dissolved in a little water and introduced into the fauces. These latter remedies under the most favorable circumstances are apt to require several hours before they operate, but they are useful in keeping up the effect. Hot-air- or vapor- or hot-baths may be used when neither pilocarpin nor jaborandi are available or efficient. Diuretics may also be given with return to consciousness. I again give the preference to digitalis, but other remedies may be substituted or alternated with it. The infusion of scoparius (broom tea) is a favorite remedy. Trousseau's diuretic wine, containing juniper, squill, digitalis, and acetate of potash, is an efficient diuretic. Squill alone I have never found to have diuretic properties. Compound tincture of juniper or gin may be used, especially where stimulants are indicated, and to meet the popular demand for them.

If *convulsions* are present, alternating with or to the exclusion of coma, *chloral* should be given. A drachm of chloral hydrate may be given to an adult by enema, often with the most promptly satisfactory results. General bloodletting should not be forgotten. The same indications which make it the best remedy in puerperal convulsions call for it here; and while it is less efficient than in puerperal eclampsia, it still may be used, with due regard to the strength of the patient.

The *asthmatic attacks*, which are a part of the uræmic condition, will be relieved by the same class of remedies, but the antispasmodics, chloral, the bromides, Hoffman's anodyne, and inhalations of ether may be used as adjuvants.

Apoplexy, which is not an infrequent termination of the disease, in consequence of the defective character of the blood-vessel walls, is recognized by the paralysis, general or partial,—most frequently hemiplegia,—which accompanies the unconsciousness. Remedies are here generally futile, but such may be used as are indicated for apoplexy elsewhere. Bleeding, counter-irritation, and, if the patient survives the immediate accident, iodide of potassium, with a view to promoting absorption of the extravasated clot, may be used.

Hæmorrhages in other situations, as from the nose or alimentary canal, are treated by the same measures as when they occur under other circumstances.

As to *specific treatment*, or treatment directed to the removal of the interstitial overgrowth in the kidney, there is only one remedy which theoretically could be expected to be of service, and that is iodide of potassium. Unfortunately the peculiar requirements of its administration, viz., the length of time during which the patient must take the remedy before any results may be expected, and the consequent difficulty in accumulating a sufficient number of cases, are such that it is almost impossible to determine whether it can be of any service or not.

Owing to these difficulties I cannot say that I have had suitable opportunities for testing this remedy. Bartels says of his experience: "The patients whom I treated upon this principle before the occurrence of any threatening symptoms—before their exhibiting any trace of dropsy or uræmia, withdrew themselves from my observation too early, so that I cannot, as yet, venture to form an estimate of the ultimate results of this method of treatment." He states, moreover, that he gives the iodide in solution to the extent of 1.5 to 2 grams (20 to 30 grains daily) for an indefinite period, and has seen no prejudicial effects from its use uninterruptedly for many months. I have long been convinced that the iodide of potassium in ordinary doses is harmless for an indefinite period. With these facts before us, it certainly merits a trial, but it is to be remembered that nothing is to be expected unless its use is begun in the earliest stages, before dropsy, uræmia, or defects of vision have made their appearance. It is best given fasting in the morning.* Smaller doses thus suffice, and there is less interference with the administration of food and other remedies required. From 10 to 20 grains a day thus administered may be continued indefinitely.

---

* The value of this method of administration, also recommended by Bartels, is well illustrated in an excellent paper published in the Philadelphia Medical Times, vol. x, 1880, p. 445, by Dr. John Guiteras, on the therapeutic advantages of administering iodide of potassium during fasting, with some remarks upon interstitial hepatitis with enlargement of the liver.

SECTION IX.

RETINITIS IN BRIGHT'S DISEASE.

By Wm. F. Norris, A.M., M.D., Clinical Professor of Ophthalmology in the University of Pennsylvania.

THE occasional occurrence of amblyopia in the course of Bright's disease has been observed ever since this malady has been diagnosticated and studied; and although Bright himself, in his first paper on the subject (1827[*]) gave no eye symptoms beyond œdema of the eyelids, he in 1836[†] recorded eleven cases, "illustrating some of the more insidious attacks which attend a fatal termination," in four of which failure of vision, coming on at a period varying from six weeks to six days before the fatal termination, was one of the prominent symptoms. In 1840[‡] the same author published twenty-four cases, with defective vision in three of them, and in 1843[§] Bright and Barlow reported thirty-seven cases of the disease accompanied by uræmic poisoning, with five cases of defective vision among them. It was not, however, till Türck[||] in 1850 demonstrated the occurrence of splotches of fatty degeneration in the retina, that the anatomical changes accompanying it were known; and six years later we find a more exhaustive study of the same subject by Virchow.[¶] The first ophthalmoscopic description of these changes was published by Heymann[**] in 1856 —just five years after the discovery of the ophthalmoscope by Helmholtz, and in 1859 Liebreich[††] added still further to our

---

[*] Reports of Medical Cases, by Richard Bright, M.D., F.R.S., vol. i, London, 1827.

[†] Guy's Hospital Reports, 1836, pp. 338-380.

[‡] Guy's Hospital Reports, 1840.   [§] Guy's Hospital Reports, 1843.

[||] Zeitschrift der Gesellschaft der Wiener Aertze, 1850.

[¶] Virchow's Archiv, vol. x, pp. 170-193.

[**] Archiv f. Ophthalmologie, Bd. ii, part 2, pp. 137-150.

[††] Ibid., Bd. v, part 2, pp. 265-268.

knowledge of the subject, illustrating his description with a colored lithographic plate. Since that period the literature of the subject has grown to such proportions that the mere enumeration of the various articles and works on the subject would require of itself many pages, and the reader will find at the end of the chapter a reference to several works where an extended bibliography is given.

Symptoms and Description.

The retinitis of Bright's disease presents very various aspects, not only in different cases but also in different stages of its development in the same case, and distinguishes itself mainly from other forms of inflammation of this nervous sheet by its marked tendency to fatty degeneration. As seen by a specialist or at an eye hospital the disease usually presents a type quite different to that predominating in the wards of a general hospital. In the former the blood-poisoning seems to fall with peculiar intensity on the nervous system, and the patients come complaining of headache, dizziness, and dim vision, these being the only marked symptoms of the malady, while the anæmia, dropsy, and other symptoms are either absent or present in so slight a degree that the patients have not supposed themselves suffering from any constitutional malady or needing advice from their medical attendant. The retinal changes in such cases are usually very extensive, and those in the cerebrum would possibly be found equally developed if we had only as accurate a method of investigating them. In the wards of a general hospital, however, we have a much better opportunity to study the development of the retinitis, and it is there most frequently encountered among those suffering from marked dropsy and cardiac disease, with transparent waxy skins, whose appearance indicates at a glance how seriously their nutrition has been impaired by the ravages of the disease. In these we often see only a few white splotches in the retina, either with or without hæmorrhages, and occasionally only a slight atrophy of the optic disk due to a previous retinitis. When the individual lives and is not markedly relieved by the rest and

13

treatment adopted, we frequently have an opportunity of seeing the development to a greater or less degree of the typical form of the affection.

Typical Cases.

In typical cases the retinal changes commence with slight œdema of the disk and surrounding retina, with a few irregular white splotches and striated hæmorrhages in the fibre-layer. We see the white patches multiply and extend, mostly within an area of two or three disk diameters from the optic entrance, until in high grades of the affection they coalesce and form a broad zone around the disk, which is itself swollen and prominent, its outlines being hidden by the opaque nerve-fibres which diverge from it.   Fresh hæmorrhages occur from time to time, and are striated when in the fibre-layer, and of irregularly rounded outline when invading the deeper portions of the retina.   They are usually either entirely absorbed or leave behind them a fatty clot, which adds an additional white splotch to those already existing in the retina.   At times they leave spots of black granular pigment as the marks of their previous presence.   At the same time irregularly linear or quadrate white splotches are developed, which radiate from the fovea centralis throughout the macular region.   These were formerly supposed when present to be absolutely characteristic of the disease, but it is now asserted by several good observers that similar appearances have been seen in the neuro-retinitis caused by brain tumor or by basilar meningitis, where there was no accompanying disease of the kidney. (Gräfe, A. f. O., xii, 2 ; Schmidt and Wegner, A. f. O., xv, 3 ; Magnus, Ophth. Atlas, Taf. vi, fig. 2 ; Leber in Gräfe and Saemisch, Bd. v, p. 581 ; Carter, Diseases of Eye, p. 382, Am. edit. ; H. Eales, Birmingham Med. Review, January, 1880, p. 47.)

Changes in the Color of the Fundus and of the Retinal Blood Columns.

In many cases occurring in the last stages of the disease a remarkably yellowish tint of the fundus is observed, together with decided alteration in the color of the blood columns in the retinal bloodvessels; the blood in the arteries being too yellow, and that in the veins presenting too little of its usually pronounced red-purple tint; in short, a state of affairs approximating in some degree to that which we find in cases of pernicious anæmia.

Curability.

Where a patient with albuminuric retinitis is admitted to hospital, and under the treatment adopted, the general health for the time improves, we not infrequently see the vision improve with it, and the retinal changes become regressive and partially disappear. I have several times seen patients with a vision of only 20–CC on admission, in a month improve so that they had a vision of 20–L, and could once more read ordinary print and perform for a time their usual duties. I have, however, never witnessed a complete disappearance of all retinal changes except in acute cases, and can well recall a case of albuminuric retinitis occurring during pregnancy, where, three years later, there was no trace of the disease beyond a slight atrophy of the optic disk.

Forms of Kidney Disease in which it may be Developed.

This form of retinitis may be developed (as has been abundantly proved by careful autopsies) during any form of Bright's disease, either with the enlarged mottled kidney of acute parenchymatous nephritis, or with the large white kidney, the amyloid kidney, or the cirrhotic kidney of chronic disease. In the vast majority of cases, however, it is developed in the later stages of the last-named form of disease, and seems to be in some way related to the blood-poisoning caused by it.

Exceptional Forms of Albuminuric Retinitis.

While the foregoing description gives a fair picture of the development of the disease as ordinarily met with, nevertheless we may encounter other varieties. Thus I have seen cases which, at the start, could not be diagnosticated by the ophthalmoscope from cases of retinal hæmorrhage from other causes (and Magnus has published similar cases), and there are recorded other cases where the only changes seen in the fundus oculi were a pronounced choking of the disk similar to that with which we are familiar in cases of brain tumor. (Magnus, Samelsohn, Gowers.)

Morbid Anatomy.

We find serous swelling of the disk and surrounding retina, especially of the neuroglia; and in the fibre-layer nests of what are usually described as varicose or sclerotic hypertrophy of the nerve-fibres. These swollen fibres are, when slightly developed, spindle-shaped, at other times so dilated that, with their large nuclei, they much resemble ganglion-cells, and they were described by Virchow as sclerosed ganglion-cells. Owing to their position, which is frequently just below the membrana limitans interna, and to the fact that their processes can readily be demonstrated to be continuous with the nerve-fibres of the fibre-layer, they are now classed by most authors as varicose hypertrophies of the nerve-fibres. This affection is, however, by no means peculiar to this disease, but is not infrequently developed in other forms of neuro-retinitis.

We find also fatty splotches, consisting of large numbers of compound granule-corpuscles, usually either in the nerve-fibre layer or in one of the nuclear layers. The radial connective tissue fibres (fibres of Müller) also present numerous minute fat-drops, which, when massed at their upper end, cause the well-known radiating white stripes on the macular region.

The bloodvessels of the retina exhibit dilatation of the veins and capillaries, with fatty degeneration of their walls, often only of their adventitia, and also so-called sclerosis, a transparent thickening of their walls, which resembles amyloid degeneration, but does not respond to the action of iodine. Hæmorrhages are almost invariably present, either striated, and more or less linear in shape between the retinal fibres, or in less regular masses in the outer retinal layers, or in the vitreous humor. The choroid often exhibits changes in its bloodvessels similar to those described in the retina, especially sclerosis of its capillaries, with fatty degeneration of their endothelium. As has been already mentioned, these changes in the retina are apt to be developed in the later stages of chronic disease of the kidneys, and therefore correspond to the period at which we find marked cardiac hypertrophy, and it has been supposed that this was essential to their development. Thus Traube claimed that the hæmorrhages arose solely from this cause, while he inclined to the belief that the other changes were due to the retention of urea. It is evident that the greater the force of the blood-current the more readily would it rupture the degenerated bloodvessels, and we must therefore assign to hypertrophy of the heart an important rôle in the production of retinal hæmorrhages ; but the fact that they are found in acute cases where no cardiac hypertrophy has had time to develop, and that many autopsies are on record where there was no cardiac hypertrophy, proves that it is not an essential factor of their production. The fact also of the abundance of retinal hæmorrhages in cases of pernicious anæmia where there is no increase of the blood-pressure, points to the same conclusion.

Statistics.

The proportion of cases of Bright's disease in which albuminuric retinitis is developed has been variously stated by different authors. Bright himself has not reported his cases with a view to determining this point, but the clinical record of them is so complete that we can readily do so. Thus we find in the four papers previously referred to records of 95 cases, in 12 of which failure of vision is recorded (12.62 per cent.). In these as in all other preophthalmoscopic data we are fairly entitled to consider the real ratio as far higher than that reported, because marked retinal changes frequently exist without decided impairment of vision, which is only necessarily interfered with when the region of the yellow spot is attacked. Frerichs[*] gives 10 cases of retinitis in a total of 78 cases, a per-

---

[*] Frerichs, quoted by Leber, in Gräfe and Saemisch, vol. v, p. 585.

centage of 12.82. Galezowski[*] gives 47 out of 154 cases (30.15 per cent.), and quotes Lecorché as giving 62 cases out of 286, or 21.71 per cent. Wagner[†] gives 12 out of 157 cases, or 7.64 per cent. He rejects, however, six cases of retinal changes, which, in his judgment, were not sufficiently characteristic, and which would, if accepted, raise the percentage to 11.46. Mr. H. Eales,[‡] the most recent writer who has examined any considerable number of cases with a view to determining the frequency of retinal disease in them, gives 100 cases, in which 28 had retinal changes, and 3 alterations of the optic disks. I regret that my own researches throw so little light on this subject; but although I have carefully examined a considerable number of cases of albuminuric retinitis, I at first devoted my attention exclusively to the mode of development of the affection, its various forms, and the kind of kidney disease accompanying them, and failed properly to note the negative cases in which there were no retinal changes. For an opportunity for studying these cases, I am much indebted to my friend Dr. J. H. Hutchinson, who has for years placed his wards at my disposal, and also for similar courtesies to Drs. W. Pepper, J. Tyson, and J. M. Da Costa. In the last 41 cases which I have examined, taken as they occurred in the wards of general hospitals, I found 11 cases of decided retinal change, such as œdema with white splotches and hæmorrhages, and have not counted several cases of slight degrees of atrophy of the disk possibly due to foregoing changes. This would give 26.82 per cent; but the number is too small to allow proper conclusions to be drawn from it, and can only be of value as a contribution to the total statistics of the subject.

Treatment.

Rest in bed with moderate purgatives and diaphoretics; in short, the remedies usually prescribed for the constitutional

* Galezowski, L'Union Médicale, 1873, pp. 924-928.
† Wagner, Virchow's Archiv, Bd. xii, pp. 219-271.
‡ Eales, Birmingham Medical Review, January, 1880, pp. 35-52.

treatment of the disease are the most effective means of clearing up such cloudy retinæ, and I know of no drug or local application to the eye which acts in any degree as a specific.

Uræmic Amaurosis.

Uræmic amaurosis is much more rarely encountered than is albuminuric retinitis in the course of Bright's disease. It is rapid in its development and in its subsidence, is without retinal changes, and the blindness is evidently due to some transient affection of the cerebral centres. It is, however, occasionally developed in cases in which albuminuric retinitis already exists.

Bibliography.

By consulting the following-named books the reader can find a full discussion of the subject and an extended bibliography, while the atlases mentioned will afford good pictorial representations of the various forms of the disease:

Allbutt, Use of the Ophthalmoscope in Diseases of the Nervous System and of the Kidneys.   London, 1871.

Leber, in Gräfe and Saemisch's Handbuch der Augenheilkunde, vol. v.   Leipzig, 1877.

Förster, in Gräfe and Saemisch's Handbuch der Augenheilkunde, vol. vii.   Leipzig, 1877.

Gowers, A Manual and Atlas of Medical Ophthalmoscopy. London, 1879.

Liebreich, Atlas d'Ophthalmoscopie.   Paris, 1863.

Jaeger, Ophthalmoscopischer Hand Atlas. Wien, 1869.

Magnus, Die Albuminurie in ihren Ophthalmoscopischen Erscheinungen.   Leipzig, 1873.

SECTION X.

SUPPURATIVE INTERSTITIAL NEPHRITIS.

Synonyms.—Pyelo-nephritis ; interstitial suppurative nephritis ; surgical kidney.

SUPPURATIVE INTERSTITIAL NEPHRITIS IS THAT FORM OF NEPHRITIS WHICH RESULTS IN THE FORMATION OF AB-SCESSES RANGING IN SIZE FROM THAT OF A MERE POINT TO THAT IN WHICH THE ENTIRE KIDNEY IS CONVERTED INTO A SINGLE PURULENT SAC.

THE term interstitial is used for this condition because the inflammation which results in these abscesses always begins in the interstitial tissue of the kidney. In contradistinction to the form of disease under consideration, the contracted or cir-rhotic kidney is the result of a *non*-suppurative interstitial nephritis. Most frequently this form of nephritis starts in the pelvis of the kidney, as a pyelitis, and thence extends along the interstitial tissue of the organ into its parenchyma. Hence the term pyelo-nephritis is often appropriately applied to the same condition. It also happens that the nephritis does not start from the pelvis, but in the interstitial tissue of the sub-stance of the organ, as when the result of an infectious embo-lus or traumatic cause, or obstruction of the tubules by con-cretions. But I do not consider it necessary on this account to make two divisions of the subject, as some authors do, con-sidering one as pyelo-nephritis, and the other as suppurative nephritis, including in the latter those cases which begin in the substance of the organ. The processes are essentially the same.

Etiology.

Probably the most frequent cause of suppurative nephritis is *retention of decomposed urine.* Such retention and decomposition are the result of long-continued obstruction to the descent of the urine from any cause, as stone in the bladder, inflammation of the bladder as the result of stone or other cause, impaction of stone in the ureter or pelvis of the kidney, or stricture of the urethra.

Calculous concretions in the substance of the kidney are another frequent cause, producing most of the cases not the result of the backing of retained urine on the pelvis of the kidney. It is easily understood how these may act as irritants to the interstitial tissue, and excite therein suppuration.

Infectious emboli cause a small number of cases of suppurative nephritis. These emboli are usually derived from the valves of the heart in cases of ulcerative endocarditis, but they may also arise in putrid wounds, stumps, or other seats of putrid inflammation. The abscesses found in the kidney in common with other organs in pyæmia are thus produced.

Traumatic injuries, as blows, kicks, or penetrating wounds in the neighborhood of the kidney, or falls from a distance and striking upon the sharp edge of a fence or similar object, may also cause suppurative nephritis. These causes, however, rarely produce the condition, because if extremely severe, rupture of the organ and peritonitis generally cause death before suppuration sets in, and if less serious they are apt to be followed by hæmaturia, which lasts for a few days, when the patient recovers.

Infectious vegetable organisms—bacteria—are claimed to be a cause of suppurative nephritis. Indeed according to some, notably Klebs,* they are an indispensable prerequisite in all cases. Neither position can be admitted as proven, but sufficient evidence has been adduced, both from pathological anatomy and clinical observation, to demand the precaution of using only scrupulously clean instruments, catheters, sounds, etc. It is very likely, at least, that decomposition in urine previously pure may be induced by these organisms, and thus an efficient cause of the pyelitis and nephritis produced.

---

* Klebs, Dr. E., Handbuch der Pathologischen Anatomie, dritte Lieferung, Berlin, 1870, s. 655.

Suppurative nephritis may occur at any age subject to the operation of the cause. The youngest patient I have met is the boy, whose case is related on pp. 208–212. Suppurative nephritis was undoubtedly here present when the boy was two years old.

Morbid Anatomy.

The appearances vary necessarily with the stage of the disease, and also somewhat with the cause.

In the earlier stage, as most frequently caused, by the retention of decomposing ammoniacal urine, which produces a diffuse inflammation of the connective tissue, starting from the pelvis of the kidney, the organ is enlarged and vascular. The capsule strips off easily, but in places drags shreds of the renal substance with it, while its under surface is otherwise slightly rougher than in health. The shreds are found to correspond with the seats of little punctate abscesses about a millimeter in diameter, on the surface of the kidney, which are thus ruptured. Others are noticed scattered over the reddened surface of the organ. They may be isolated or arranged in groups of from three to six. At this stage, too, they will be found scattered over one or more areas, each corresponding to the base of a single Malpighian pyramid, while other pyramids may be entirely free. Each abscess is surrounded by an intensely red border. On principal section of the kidney, these little abscesses appear as yellow striæ running in the direction of the tubules, between which they are found on minute examination to be placed. The striæ converge a variable distance into the medulla, becoming, as they do so, more closely aggregated, and extending occasionally as far as the papillæ. In the medulla they accompany the connective tissue about the vasa recta, and in the cortex the interfascicular veins, the beginning of which on the surface of the kidney corresponds with the punctate abscess, which latter is, as it were, the outer end of the streak. According to Klebs they contain, in addition to pus-corpuscles, minute shining granules, which are not altered by the action of alkalies or acids, are dissolved neither by alcohol nor ether, and which he considers, therefore, bacteria.

At a later stage, these little collections of pus unite to form larger ones, these again to form others still larger, destroying the tubular structure of the kidney as they encroach upon it; and it is at this stage that cases of pyelo-nephritis not unfrequently terminate unfavorably, and the specimens come under observation. At first each of the abscesses thus formed is confined to the region of a single pyramid, and it not unfrequently happens that a kidney is partitioned off in the manner shown in Fig. 33, drawn from a specimen in the possession of the author. Before this occurs, however, the abscess bursts through the papilla and calyx into the pelvis of the kidney. The pelvic end of the papilla is then bounded by an uneven ulcer, which gradually enlarges and deepens until the entire pyramid is destroyed, with more or less of the cortex corresponding to it. If the case lasts long enough even the partitions referred to may be eroded, and thus it occasionally happens that the entire kidney is converted into a huge purulent sac. This of course can occur with one kidney only, while the other is, in a measure, able to continue its function of secretion, although the two kidneys are sometimes so far altered that it seems incredible that death from uræmia should not have occurred much earlier in the course of the disease.

More rarely, it happens that the abscess ruptures through the capsule and leads to the formation of subperitoneal abscesses, which may open posteriorly or gravitate towards the pelvis and open under Poupart's ligament.

The pelvis of the kidney is always dilated, and the seat of a purulent catarrh, the product of which passes down the ureter with the urine into the bladder. As often, the ureter is similarly dilated, being sometimes converted into an intestine-like tube. This is more particularly the case when it is impacted with the calculi, as in the drawing presented, or there is some decided obstruction to the passage of the urine from the ureter into the bladder. The pelvis of the kidney itself is even more frequently impacted with calculi than the ureter, when the latter, unless it at the same time contains stones or is also obstructed at its outlet, is less dilated. According to the degree of obstruction, the pelvis and ureter also contain more or less decomposing and stinking urine, mixed with pus and crowded with bacteria.

Almost invariably fatal as these cases are, a complete arrest of the process is said to be not impossible, followed by inspissation, caseation, calcareous metamorphosis, encapsulation, and recovery.

Klebs* describes a still earlier stage of pyelo-nephritis than that of the punctate abscesses described. In this, the kidney is enlarged and vascular and the capsule non-adherent. He says that at this stage there are no *inter*tubular changes whatever; that the tubules themselves are broad, their epithelium cloudy, sometimes fatty ; their lumen is widened and *filled with bacteria.*

Ebstein, in his recent article in Ziemssen's *Cyclopædia of Medicine*, vol. xv,† confirms the description of Klebs, and sustains his view as to its parasitic origin, while he also claims that it has now the most adherents. These adherents contend that the bacteria themselves excite the inflammation, causing, in the first place, a purulent pyelitis, and subsequently circumscribed renal inflammation.

They adopt the suggestion of Traube, that the bacteria gain admission from without, generally by the introduction of dirty catheters. I regret that I have nothing to offer from experience of my own in the examination of kidneys in this early stage of suppurative nephritis, but supposing the observations of Klebs and Ebstein to be correct, the question of the parasitic origin of pyelo-nephritis becomes only a part of that of the parasitic origin of disease generally, which, to say the most, must be considered *not proven.* On the other hand, we have to remember, first, that numerous cases of pyelo-nephritis have occurred in which a catheter has never been introduced ; and, second, that very few occur in which there does not exist a cause in the shape of a mechanical irritant which is amply sufficient to account for the irritation.

---

* Klebs, op. citat.				† William Wood & Co., New York, 1877.

Where the abscess is *embolic* in origin its seat is at first occupied by an area of intense hyperæmia, resulting in hæmorrhagic extravasation, which takes place also into the tubules, causing bloody urine. To this succeeds suppuration. The size and number of the abscesses depend upon that of the plug obstructing the bloodvessel, which is usually one of the interlobular arteries or a vas afferens. The embolic abscesses may also be multiple in consequence of the breaking of the embolus into a number of minute fragments.

Where the cause is *traumatic* the process is not so easily defined. Circumscribed abscesses may occur, or the kidney may be converted into a soft pulpy mass, a mixture of pus, blood, and broken-down renal substance.

Symptoms and Course of Suppurative Nephritis.

The symptoms of this condition are not numerous, and, apart from the characters of the urine, are not very distinctive. *Pain and tenderness* are the most constant, but it has occurred that considerable inroads have been made upon the structure of the kidney without pain resulting. On the other hand, the pain is often of a very severe character, while the tenderness over the region of the kidney affected is also evident. Most frequently, but not always, the severest pain is in the region of the kidney itself, whence it radiates towards the front of the abdomen and the groin, and is accompanied often by retraction of the testicles. Where the condition is the result of impacted calculus, the seat of the impaction is the seat of pain. Thus, in the little patient whose case affords a typical illustration of one due to impacted' calculi, and which is therefore narrated at the end of this section, the region between the umbilicus and the pubis was the seat of numerous and severe paroxysms of pain, and the post-mortem examination revealed that both ureters, in that part of their tract corresponding to this region,—on the lumbar vertebræ,—were thoroughly impacted with calculi. The pain is always intermittent as to degree, sometimes totally so, but generally there is more or less constant pain of a less degree, which is paroxysmally in-

creased. Various positions are assumed by the patient with a
view to easing the pain, among which that on the face is not
infrequent.

A distinct *tumor* can sometimes be discovered in the region
of the kidney by palpation and percussion. This implies an
enlargement of the organ, which is either due to its complete
conversion into a purulent sac, or an augmentation of its size,
due to the distension of its pelvis with pus or calculi or both.
In the case to be narrated at the end of this section a distinct
tumor could be felt below the umbilicus, which was found, at
the post-mortem examination, to be due to a mass of calculi
impacted in both ureters, in this locality.

Fever is also an intermittent symptom. Possibly in a very
few latent cases it may be altogether absent, but except in these
there is always slight frequency of pulse and slight elevation
of temperature. These latter at times become decided, and in
advanced stages the fever is sometimes hectic, being followed
by profuse sweats.

In acute cases the beginning of suppuration is often marked
by a *chill* or succession of chills, but in most instances it is quite
impossible to recognize the beginning of the suppurative stage.

The *characters of the urine*, as intimated, are more distinctive. This, except in acute infectious cases (pyæmia), almost
invariably, sooner or later, contains *pus*, and unless it does
contain pus no certain diagnosis can be made. *Blood* is also
a very constant constituent of the urine from cases of suppurative nephritis, but while such urine is scarcely ever examined
by the microscope without discovering a few blood-disks, yet
the quantity is not often large enough to be recognizable to
the naked eye. The quantity of pus is also very variable.
While it may be so copious as to produce a heavy white opaque
deposit, equalling one-sixth to one-fifth the bulk of urine, it
may be represented by little more than the normal proportion
of leucocytes. This variation will also occur at different times
in the same case. Pus from the kidney and its pelvis is distinguished from that formed in the bladder by the absence of
that glairy property which is so characteristic of the latter, due

to the admixture with mucus and the alteration which the pus itself assumes by the action of the carbonate of ammonium formed out of urea during decomposition.   Pus from the pelvis of the kidney is also occasionally fetid, containing bacteria, and prone to decomposition, but very rarely as compared with pus from the bladder.

Tube casts, on the other hand, in my experience, are very rare in this form of kidney disease.

The urine is also diminished in quantity, the degree of diminution depending upon the proportion of kidney structure destroyed in the suppurative process.   Complete suppression is not uncommon towards the close of cases presenting extreme degrees of destruction. Notwithstanding such diminution, the color is pale and the specific gravity low, owing to the small proportion of urea present.   I have known the range of specific gravity in a single case to be from 1003 to 1016.   In *reaction* the urine is faintly acid, neutral, or alkaline, and, as already stated, is often prone to rapid decomposition, and therefore to assume the alkaline reaction.

It *is always albuminous*, but the quantity of albumen is never excessive, and varies generally *pari passu* with the quantity of pus and blood.   Yet I have been much impressed with some disproportion in this respect, to which I am not aware that attention has heretofore been called.   Thus I have noticed in some cases a disproportionately large amount of albumen associated with a small amount of pus, and, on the other hand, a quite large bulk of pus is sometimes associated with but a trace of albumen.   I am as yet unable to explain these discrepancies, but think I have found the degree of structural disintegration greater in those cases where there was a disproportionately large amount of albumen, as compared with the quantity of pus.   Such cases are, therefore, more serious.   It sometimes happens that there is a sudden increase in the quantity of pus in the urine, followed by a gradual diminution, or the urine previously clear may suddenly become loaded with pus.   Such occurrences indicate the probable period of rupture of an abscess through a papilla, and a pouring out of its contents into the pelvis of the kidney. Accumulations and sudden evacuations of this kind may also be due to temporary obstructions to the descent of the pus. It rarely happens that a small portion of the substance of the kidney is thus discharged with the urine, when it may be recognized by microscopic examination, which will discover the tubules and Malpighian bodies of the kidney. Two instances of this are related by Ebstein.*

Occasionally, also, the abscess, instead of rupturing into the pelvis of the kidney, perforates into the perinephritic tissue, burrowing in different directions and producing fistulous openings. Perforations may thus take place posteriorly in the lumbar region, or anteriorly at the groin, especially into the colon, and more rarely into the lungs and liver, and even peritoneal sac.

The *course* and *duration* of suppurative nephritis are very various. Traumatic cases are comparatively rapid, either to recovery or death. Pyæmic cases may run their course in forty-eight hours, and are invariably fatal.† But cases due to impacted calculus, to stone in the bladder, cystitis, or other cause of obstruction to the descent of urine from the kidney, may be prolonged indefinitely, while some terminate without being discovered. Sooner or later the patient generally succumbs to exhaustion, but even in youth, life may be sustained for years with paroxysms of the severest suffering and a surprising degree of destruction of the kidneys. The greatest danger to those thus affected, is intercurrent illness, which is always more seriously influenced and much more apt to terminate unfavorably. It is well known that the operation for stone is much more apt to be followed by a fatal result when the subject happens to have a surgical kidney, a very prominent instance of which is that of the ill-fated Napoleon

---

* Ziemssen's Cyclopædia of Medicine, New York, 1877, vol. xv, p. 557.

† In an able experimental thesis, presented to the Medical Faculty of the University of Pennsylvania, March, 1881, Dr. Louis Brose has shown that interstitial changes may make their appearance in 22 hours, and abscess of the kidney at the end of six days after injecting tincture of cantharides.

III, late Emperor of the French.   The operation is, however, not necessarily fatal, even when there is suppurative nephritis of both kidneys, as is shown by the appended highly interesting and illustrative case:

On April 21st, 1877, when 3 years and 10 months old, W. H. was cut for stone by Professor D. Hayes Agnew, who removed an oval, smooth phosphatic calculus, 3.5 cm. long by 2.2 cm. in its longest conjugate diameter; on section, also white, loose in the texture of its central portion, and without a nucleus.   It was of the same composition throughout.   The cut made by the operation was very slow in healing, the latter not being completed until October 1st.   At least two years before the operation—that is, when less than two years old—he began to complain of bladder symptoms; previous to this he had been delicate.   He came under my care about May 1st, 1878, one year after the operation for stone.   He was then suffering with the symptoms of cystitis, such as are present with a calculus in the bladder, but usually subside soon after the removal of the stone.   There was frequent micturition, a small amount of pus, and a very small degree of albuminuria, while the urine was either alkaline when passed, or became so very soon afterwards.   He was subject, however, in addition, to paroxysms of extreme pain, which was confined almost exclusively to the region of the navel.   The attacks occurred about once a week, and succeeded upon days on which he felt unusually well, played a good deal, and tired himself thereby; they lasted until paregoric sufficient to relieve them had been given.   He was very pale, anæmic, and appeared a very delicate child.

I put him on benzoic acid, which was to be given until the urine was distinctly acid, and at the end of two weeks he was greatly better.   The paroxysms were less frequent; there was less pus in the urine.   To the benzoic acid treatment was added sandalwood oil, and the two were continued, more or less constantly, either together or alternately, as the condition of the urine and other symptoms suggested, throughout the entire summer.   This was spent at the seaside.   Iron and quinine were also given as tonics when required.   It was considered that during this time there was a gradual abatement of his severer symptoms.

About September 10th he had a decided return of his symptoms, this time accompanied with great pain over the bladder and frequent micturition, during which he seized his penis in his hands; he passed water from three to six times in a night.   His general health, I thought, at this time greatly improved.   The urine was very pale, almost colorless, acid, specific gravity 1003, and contained a few leucocytes and a small quantity of albumen.   By the middle of October he was again relieved of the painful symptoms, the specific gravity of the urine increased to 1007, but the albumen had increased decidedly to half the bulk of urine tested.   No tube-casts could be found, nor were any ever found in the entire course of the sickness, at least while he was under my observation.

14

By the middle of January, 1879, the albumen had again become very small in quantity, there was a scanty sediment of leucocytes, and the urine had attained a specific gravity of 1016,—the highest it ever was while he was under my care. The special treatment during this time was sandal-wood oil directed to the cystitis, and benzoic acid, to influence the reaction of the urine, as required.

On February 3d, 1879, he had a hæmorrhage from the bladder, accompanied by great pain. It was, however, at a single act of micturition, and what I examined by the microscope, eight hours after this, contained no blood-corpuscles, but one-half its bulk of albumen. The hæmorrhage did not recur.

February 12th, Professor Agnew sounded for stone successfully. On March 11th he operated, and, after some difficulty, removed the second stone. This was much smaller than the first, being also oval, 2 cm. long by .6 cm. wide. More than half the external surface was white, the remainder brown, and projecting from the former were a number of spicules, by which it was supposed it was imbedded in corresponding depressions in the bladder, whence the difficulty in extracting it. On section, the external concentric laminæ were white, but in the centre was a brown, oval nucleus, 1 cm. long and .5 cm. wide. Portions of this nucleus almost totally disappeared on incineration, but the residue responded to the murexide test. It, therefore, contained some uric acid.

He was kept in bed six weeks, until April 30th, when he was allowed to get up, although the wound was not quite healed. Until April 15th there was a good deal of pus and some blood in the urine, but upon that date we returned to the treatment by sandalwood oil, after which they rapidly diminished. He had some pain on passing water, which continued until April 15th, when this too disappeared. Immediately after getting up the urine dribbled through the opening in considerable amount, but this gradually diminished until it was scarcely sufficient to soil a napkin during the entire day. On May 10th the urine contained a mere trace of albumen, a small sediment composed of pus and earthy phosphates, was faintly acid when passed, but became alkaline soon thereafter, and presented a specific gravity of 1005. Soon after this he went to the seashore.

On August 14th the little patient's mother called upon me to report his condition. She said that for as much as two weeks at a time there was no dribbling of urine from the wound; then an attack of pain was succeeded by dribbling for a time. He was very well in every other respect, although thin. He ate heartily and played all day long. He continued taking the sandalwood oil, and for a time the benzoic acid.

Early in September the boy returned from the seaside. He was very thin, but his wound had entirely healed. His mother reported that soon after his return he took cold two or three times in rapid succession, his stomach became deranged, his appetite disappeared, and he became weak and emaciated. I saw him September 20th, the first time since his return, and was much struck by his emaciated and cachectic appearance. The urine presented about the same characters as when last examined. He was immediately put to bed, placed on restorative treatment, and for a time seemed to grow stronger; but, as though in consequence of the absence of treatment especially directed to the urine, the latter contained much more mucus and became alkaline, and these characters continued until the benzoic acid was re-ordered, when they partially disappeared.

The improvement in his general condition was, however, only temporary, and early in November it was evident he was again declining. He emaciated, grew weaker day by day, and died on the afternoon of November 15th, 1879, although he was out in his coach two days before. On the day

FIG. 33.

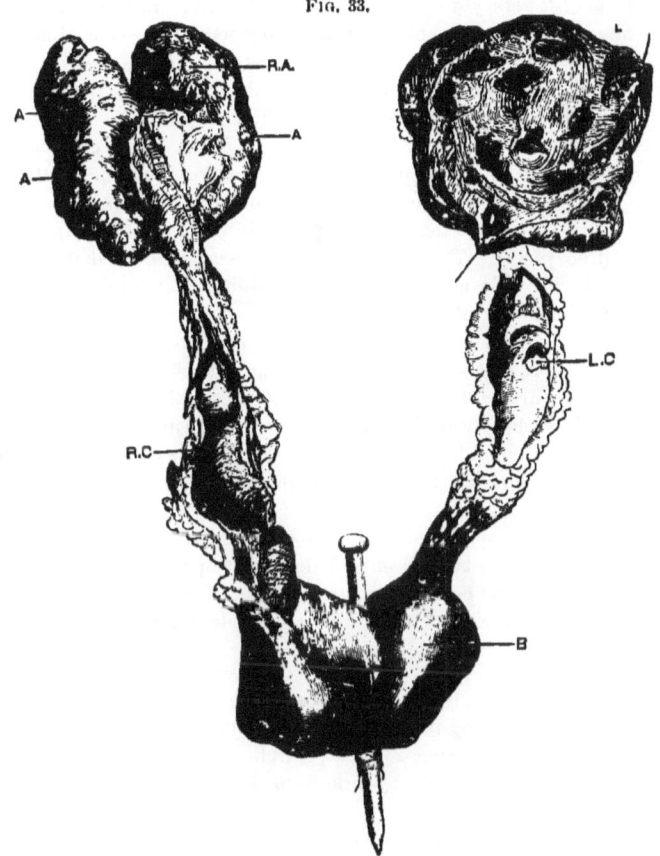

before his death he had to be catheterized, but there was considerable urine in his bladder, and there was no reason to suppose there was suppression of the secretion. He also suffered a great deal of abdominal pain for two or three days before he died. Very interesting in connection with the results of the post-mortem examination is the fact that his father, while feeling his abdomen on the afternoon before he died, discovered a hard, irregular mass in the neighborhood of the umbilicus.

The *post-mortem examination* was made fifty hours after death. The body was extremely emaciated, but all the organs were found normal except the kidneys, ureters, and bladder. Both kidneys were cystic, each being converted into a multilocular cyst, of which the loculi were distended with a mixture of pus and urine, the former apparently predominating, as there was very little urinous odor. The left kidney was the larger, measuring 10.5 cm. by 5.5 cm. Its surface contained numerous white tubercles (see A, A in the figure), ranging in size from that of a pin's head to a pea. The *pelvis* of this *left kidney* was distended with the same fluid which filled the loculi, and the left ureter—dilated throughout its extent—was almost completely occluded by a single calculus of most peculiar shape. It was spirally twisted and somewhat spindle-shaped, much constricted near its middle, and 5.5 cm. long by 1.5 cm. in its thickest part.

The *right kidney* was smaller, measuring 8.5 cm. by 4 cm., but in all other respects similar. The pelvis and ureter were dilated, and the latter packed with three fragments of what was originally a single stone, constricted at two points, where it had probably been fractured in removal, making three pieces. The original stone was arcuate in shape, 10.5 cm. long and 1 cm. wide at its middle. The same tubercles were scattered throughout the cortex of this kidney.

The *bladder* contained a small stone, which had evidently descended since the last operation, and was apparently a fragment of one of the stones remaining in the ureter, the fracture having taken place at a point of constriction. This stone was about 2.5 cm. long, and varied in diameter from .5 to .75 cm. The bladder was thickened, the mucous membrane thickened, and no traces of the cut made at the time of the operation were discoverable.

The pediculation observed on the stone removed at the last operation, which was believed to have occurred during its presence in the bladder, most probably did not occur there, as the stones found in the ureter exhibited the same peculiarities, although these pedicles were probably the means of its attachment to the bladder.

This case was a rarely interesting one, not only in its unique morbid anatomy, and as showing the rapidity and extent of sedimentary processes in the urine, but also in suggesting a possible explanation of certain recurrences of stone in the bladder; while from the therapeutic aspect it is not without its uses.

It is evident that the hard mass felt through the abdominal walls just before death was made up by the calculi, which, although contained in separate ureters, united to form a single confused mass in the median line just above the sacrum. How long before death this could have been felt I do not know, for I had not examined the abdomen for some time. There is every reason to believe, however, that it might have been recognized at any time subsequent to the emaciation, which began during his absence from the city, in the summer, and which so impressed me when I first saw him, some time after his return. The practical hint to be deduced from it is that frequent examinations of the abdomen should be made in cases of suspected impacted calculus.

There are no *complications* peculiar to suppurative interstitial nephritis other than those mentioned as causing it or as resulting from unusual accidents of rupture, of abscess, and perforation of neighboring organs.

Diagnosis.

The diagnosis of suppurative nephritis may be easy or difficult. It is easy when there is the history of a traumatic cause followed by hæmaturia, and later, purulent urine, with tenderness and pain over the region of the kidney. On the other hand, while the long continuance of inflammation of the bladder, stone in the bladder, nephrolithiasis, or other causes of decomposition of urine and of obstruction to its descent, are always just causes of suspicion that suppurative nephritis exists, it is not easy to find data on which to base a positive assertion of its presence. If, however, the urine contains pus which by its characters is known to come from a source above the bladder, either permanently or at intervals, along with the symptoms of any one of the conditions named as causing suppurative nephritis, its existence may be averred with tolerable certainty. And if, moreover, there is tenderness over the region of the kidney, and especially if a tumor may be mapped out, there is less doubt. Beyond doubt this category of cases —those due to obstruction—is by far the most important to the practical physician, including, as it does, those which he most frequently meets.

A most important practical question, one which if correctly answered would be of infinite service in diagnosis, is the following: How long can obstructed conditions of the ureter and pelvis of the kidney exist before producing suppurative nephritis? Unfortunately I have no precise data on which to base an answer to this question. In the case of the boy Willie, whose case is related on pages 208–212, I have no doubt the condition was present when he came under my care one year after the first operation for stone. The stone was in the bladder two years before it was removed, so that the suppurative process was present within three years after the process of impaction set in, and probably sooner. I think, however, it may be safely stated that *where obstructive conditions of the ureter or pelvis have existed for three years, the corresponding kidney or kidneys are probably the seat of suppurative nephritis, and partially cystic.*

Pyæmic abscesses of the kidney can only be suspected to be present as a part of the pyæmic process, which is recognized in surgical cases presenting conditions favorable for its causation, by the occurrence of chill and other symptoms peculiar to it.

Prognosis.

The prognosis so far as recovery is concerned is unfavorable. Traumatic cases may recover if the injury is not too extensive, while very grave injuries are usually rapidly fatal. Cases due to obstruction of the ureters cannot get well as long as the obstruction and irritation continue, and as their removal is often impossible such cases gradually grow worse. On the other hand, their fatal termination may be indefinitely delayed. It is often a matter of surprise and wonder, on viewing the post-mortem appearances of cystic purulent kidneys, that the patient has lived as long with the extreme structural changes. which are found to exist, the barest remnant of secreting structure being sometimes found to be present. It is impossible to say how extensive a lesion of the kidney thus produced might be repaired, provided the removal of the cause could be secured. Conditions of this kind might occur where a stone in the bladder has been removed after having been long enough present either in the bladder itself or in the ureter to cause suppurative nephritis. While it is not likely that the essential structure of the kidney could in any degree be reproduced, there is scarcely any limit to which such as remains could be made to supplement the lost function of parts

completely destroyed. The fact remains that there is no positive evidence to prove that any one has recovered either partially or completely from a suppurating cystic kidney, while
numerous instances are constantly occurring in which the victims of this disease have lived for many years, sometimes even
unconscious of ill-health. It is scarcely necessary to say that
such persons are in imminent danger from the operation of
any cause, such as cold or acute disease, which tends to sup-
press the action of kidneys already crippled in their function.
The following case illustrates my meaning:

The patient was a gentleman, forty-three years of age, whose business
was farming and the purchase and sale of cattle, on account of which he
made frequent fatiguing journeys to the West. He first consulted me in
May, 1878, and stated that his first illness was in 1864. It was evidently
nephritic colic. Severe lumbar and abdominal pain continued to recur,
more particularly after exposure to cold, until 1865, when he passed a small
calculus. He was relieved for three years, having then another attack,
during which he passed a sediment which he described as "steel-colored,"
and, finally, another small calculus, also "steel-colored." There was again
an interval of relief until 1873, when there occurred a long series of attacks,
in the last of which, in 1874, he passed quite a large stone, three-quarters
of an inch long and one-third wide, which he described as having a white
coating (phosphatic?). From this time up to the date of his visit, there
had been more or less dull pain. His urine, which for some time previous
had been dark-hued, had lately assumed a lighter hue, and while brick-
dust sediment formerly was observed it had lately disappeared. Pus was
first observed in his urine in 1875, by Professor Reamy, of Cincinnati. He
had had no acute attack during the year 1878, up to the date of his visit to
me. He had been sounded for stone in the bladder four times—twice in
1864, once in 1865, and once in 1875—without any result.

At the time of his visit to me he was rising at night to pass water from
four to eight times, and during the day could not retain it longer than one
hour and a half. His urine was pale amber in hue, had a specific gravity
of 1012, and deposited a considerable amount of pus ($\frac{1}{15}$ its bulk). There
was also a small amount of albumen, not more than a line by Heller's acid
test. The bladder symptoms were very much relieved by sandalwood oil,
but he continued to have attacks of lumbar pain, always after exposure to
reduced temperature, especially if he happened to be overheated at the
time.

On the evening of January 15th, one week after his return from a trip
to the West, during an extremely cold season, he was seized with one of his
attacks, the pain being chiefly on the left side. The attack was one of unusual severity, but he was relieved by a hypodermic injection of half a grain

of sulphate of morphia by his physician, Dr. W. H. Barr, of Middletown, Del. The relief was complete, but he remarked to Dr. Barr, on the morning of the 17th (about thirty-six hours after the attack), that he had not passed water since two o'clock of the 15th, and had no desire to pass any. I saw him on the 18th, reaching Middletown at two o'clock, just seventy-two hours after he had passed water. The bladder was empty, as shown by the use of the catheter. He was entirely free from pain, but was suffering a good deal from nausea, which had made its appearance the day previous. He was slightly drowsy, but he was so far conscious that he could pass the catheter on himself, as was occasionally his custom. The drowsiness, however, increased during the four hours I spent with him. He was cupped over the loins, and the cupping followed by hot cataplasms, and, in the course of the next nine hours, was freely purged by elaterium, and sweated by jaborandi. The treatment availed nothing, however, in relieving the suppression. The drowsiness increased, and he died at noon of the 20th, one hundred and eighteen hours (within two hours of exactly *five* days) after he last passed water.

The *autopsy* was made by Dr. Barr, fifty-six hours after death. The bladder appeared not to have been examined, but all the other organs in the abdomen except the kidneys were normal. The latter I presented to the Pathological Society of Philadelphia, but the left one had lost almost entirely the appearance which made them, in conjunction, among the most striking specimens of morbid anatomy I have ever seen. The *right* was sacculated, the cysts ranging from half an inch to one and a half inch in diameter, and were round and oval; in one of them was imbedded a calculus as large as a pea. The others were filled with a yellowish and odorless fluid containing albumen. They were not lined with an epithelium. The kidney was also larger than the normal organ, weighing six and a half ounces. The capsule was strongly adherent. *Microscopic examinations* of thin sections exhibited the appearances of interstitial nephritis, and the condition was evidently one of long standing, the result of the constant hyperæmia caused by the calculi which have from time to time formed, and all of which seem to have escaped except the one still shown in position. In the parenchyma numerous infarctions of blood were seen, and the epithelium was fatty in places. The *left* kidney exhibited a great and most interesting contrast to the right. It was throughout of a beautiful bright-scarlet color, weighed seven ounces, and was hard in consistence. The capsule was only slightly adherent; surface smooth. On section, it presented a homogeneous, very bright-red color. There were no cysts. The pelvis was somewhat dilated. Microscopic examination showed large, swollen tufts, the epithelium of the tubules largely swollen and desquamated, the bloodvessels normal but overdistended by blood, with other features of a parenchymatous nephritis.

Treatment.

There is no curative treatment for suppurative nephritis without a removal of its cause, and as the latter is almost invariably impossible, it follows that our measures are mainly palliatives.   One of the most frequent indications is the *relief of pain*, which is usually so severe as to call for powerful anodyne measures,—*opium* and its alkaloids being absolutely essential.   Hypodermic injections of the salts of morphia, in doses of $\frac{1}{8}$ to $\frac{1}{3}$ of a grain, repeated, if necessary, are favorite and effectual methods of relieving the intense pain, which is often due, not so much to the inflammatory process as the cause of obstruction.   Suppositories of $\frac{1}{2}$ to 2 grains of the extract of opium may be substituted.   Hot fomentations are also valuable adjuvants.

The *catarrhal process* which is constant in the kidney itself, in its pelvis, and the ureter, and also in the bladder, requires treatment, and, although in consequence of the persistent operation of the cause is incurable, is nevertheless capable of decided improvement.   The only remedies I have found efficient are the balsams and benzoic acid.   Of the former I prefer *sandalwood oil*, both on account of its efficiency and because it is better borne by the stomach.   Given in gelatin capsules, each containing 10 drops, of which one or two may be taken three times a day, it will be found to have a decided effect upon the catarrhal inflammation, seen in a diminution of the amount of pus in the urine.

Benzoic acid fulfils another indication, that of securing an acid reaction of the urine, which is very often either alkaline or so faintly acid that it rapidly becomes alkaline, and thus predisposes to decomposition.   The benzoic acid is best given in the form of compressed pills.   For an adult three or four 5-grain compressed pills daily are usually sufficient to keep the urine acid.   Larger doses than these may be given.   It may be given either alone or in conjunction with the sandalwood oil, the former being given before and the latter after a meal.   To children smaller doses may be given.   I usually begin with one grain three times a day and increase it.   To the boy Willie, whose case is related on p. 208, I gave five grains three times a day.   I have never found the careful administration of these remedies to produce derangement of the stomach, and by their use have greatly ameliorated the symptoms. Benzoic acid is sometimes advantageously combined with opiates.

The various vegetable diuretics, as buchu, pareira brava, etc., I have never found of any use in these affections, suppurative nephritis and pyelitis.

The constant and inevitable tendency in these cases to run down in general health, in consequence of the drain and wear and tear to which they are subjected, renders due attention to restorative measures essential, and the use of quinine, iron, milk, and other nutritious articles of diet is constantly indicated; while the dangers to which the patient is subjected from exposure require constant precautions to be observed against cold and dampness, which, by inducing acute nephritis, may lead to a rapidly fatal issue, as illustrated by the case just related.

SECTION XI.

CYANOTIC INDURATION OF THE KIDNEY.

Synonym.—Passive congestion of the kidney.

CYANOTIC INDURATION IS A PECULIAR INDURATED STATE OF THE KIDNEY DUE TO A SIMPLE HYPERPLASIA OF ITS INTERSTITIAL TISSUE, WHICH RESULTS FROM LONG-CONTINUED PASSIVE CONGESTION.

Etiology and Pathogeny.

WHILE any agency which obstructs the movement of the blood through the kidney may become a cause of cyanotic induration, the only ones encountered in actual practice are valvular disease of the heart and chronic pulmonary diseases involving considerable areas of the lung, such as emphysema, phthisis, etc., or pleurisy, with extensive effusion, or marked adhesions, etc.

In either event the mechanism of its production is the same. The blood is crowded into the venous side of the vascular system. In mitral insufficiency the blood is regurgitated from the left ventricle into the corresponding auricle, and thence into the lungs; the latter organs become engorged, and again resist the entrance of blood from the right side of the heart, whence it is backed into the valveless vena cava. The smaller veins of the extremities at first resist it by means of the valves with which they are provided. But the veins of the abdominal organs, including the liver and the kidneys, are without valves, and are the first, therefore, to receive the brunt of the stagnation. They become gorged with blood, and it is as though a string were tied around the renal vein, preventing the exit of the blood. What is the consequence? The connective tissue so abundantly present in the liver—and although sparsely present in the kidney, still there in sufficient quantity to become the starting-point of new formations—becomes infiltrated with *liquor sanguinis*, the natural pabulum of the tissues. Thus supplied with food, the connective tissue corpuscles proliferate. Others are formed by the proliferation and fixation of the amœboid white corpuscles, which, under the favorable circumstances afforded by a stagnated circulation, wander out in great numbers from the capillaries and small veins. These new cells are differentiated into connective tissue fibres. The condition is really that of an interstitial nephritis, but for the reasons stated on p. 84, I shall consider it here briefly.

In pulmonary or pleural disease the obstruction begins in the lungs instead of the heart, but the mechanism is the same.

Morbid Anatomy.

The kidney of cyanotic induration or passive congestion is hard, firm, and bluish-red as to its external surface. In the earlier stages it is enlarged simply from the presence of the large amount of blood detained in its parenchyma. The stellate veins are unusually distinct. The capsule strips off easily, and on section the enlargement is found to involve the cortex but the veins of both cortex and medulla are engorged, that of the straight veins causing the medulla to appear darker in hue than the cortex. The Malpighian bodies, on the other hand, are not always engorged. The cut surface of the kidney is moist and succulent, but the microscope reveals no further changes, either in the cortex or the medulla, the epithelium being unchanged.

But the kidney is rarely seen in this stage. When found at autopsies of cases of heart disease, the enlargement referred to has nearly or quite disappeared, and the kidney is slightly if at all larger than the normal organ, though rarely if at all smaller. The other superficial characters of hardness, smoothness, and bluish-red color, however, remain. Sometimes there appears a slight tendency to lobulation, or even a slight disposition to unevenness, by reason of certain shallow superficial depressions, but very seldom anything like a granular appearance of the surface of the organ stripped of its capsule. In this event the capsule does not strip off quite as easily as usual, but may drag small portions of the parenchyma with it.

On section the relations of the cortex and medulla are not much altered, but the succulency of the parenchyma which characterized the early stage has been replaced by a uniform hardness throughout the entire organ. The capillaries are less turgid, and contribute less of their hue to the cortex, which is therefore paler, although the Malpighian bodies may be relatively more distinct by the retention of the blood in their interior. The microscope explains the cause of the induration in a simple rather than a corpuscular overgrowth of the connective tissue between the tubules and vessels, that is, there is no relative increase in the proportion of cells, but a uniform increase of both the cells and fibrillar-intercellular substance. This tends slightly to contract, to compress, the capillaries, and thus interfere with the proper nutrition of the cells, which may become more granular or even fatty, and may waste, resulting in destruction of portions of the convoluted tubules; hence the superficial depressions of the surface alluded to.

Occasionally there may be a corpuscular increase in the interstitial tissue, that is, there may be an increase of the lymphoid cells between the tubules; but this is rare, the process being in the main one of simple hypernutrition resulting from a copious access of pabulum rather than a process of an active inflammatory character.

The condition is also to be distinguished from a pre-existing granular kidney, upon which cardiac disease may have been ingrafted secondarily.

Symptoms of Cyanotic Induration.

The symptoms of this condition are primarily those of the disease of which it is the consequence, which I will not repeat.

To these are superadded generally a dropsy, scanty urine, of high specific gravity, containing usually a small amount of albumen and a few small hyaline casts.

The *dropsy* is usually of the lower extremities, in the area drained by the inferior vena cava, while it will be remembered that that of renal disease more frequently involves the upper extremities. There also occur, however, effusions into the pleural sac and peritoneum, and the hands and arms may be involved.

The *urine*, as stated, is scanty, and of high specific gravity, often 1030, and even higher. It is turbid with urates, depositing a copious sediment of them and of uric acid.

The *albumen*, as stated, is usually small in quantity, but may become larger if the obstruction to the movement of the blood is great.

The *casts* are small, transparent, or faintly granular, and not numerous, indeed often absent. Fatty casts are also occasionally present.

Further, a kidney thus constantly engorged is much more liable to attacks of acute inflammation than one which is in the normal state. This fact explains the frequency with which, in cases of this nature, the urine becomes suddenly bloody, highly albuminous, and filled with epithelial and blood casts, even after slight exposure, such as would be without effect upon a healthy individual. In the intervals between such attacks the blood-casts disappear entirely, and the epithelial and granular casts also become much less numerous, and often disappear, the granular continuing longest, while even fatty casts which were present may disappear.

Diagnosis.

The supervention of renal disease upon cardiac disease may be suspected when all the symptoms of the latter become aggravated, the dropsy becomes peculiarly persistent, and the urine is scanty. The high specific gravity of the urine, and the presence of albumen and casts, confirm the suspicion.

Prognosis.

With the supervention of the renal involvement, the inconveniences and annoyances of the cardiac disease become many times greater, while the difficulties in the way of improvement are nearly proportionately increased. Yet the results which sometimes follow appropriate and energetic treatment, and the substitution of favorable for unfavorable hygienic surroundings, such as succeed the admission of a neglected outcast to the wards of a hospital, are often astonishing. Under these circumstances it is not unusual for the dropsy to decline, the albumen and casts to disappear, and the patient to be restored to comparative comfort, without, however, any change in the original lesion, which upon the slightest provocation will re-excite all the symptoms.

Treatment.

As intimated under prognosis, the substitution of favorable for unfavorable hygienic surroundings is the primary requisite. Shelter, warmth, rest, good food are indispensable. After this *digitalis* is our most powerful lever, and for evident reasons. We have here to deal with a dilated, weak, failing heart, unable to drive the blood forward. Its power must be increased, and we have a remedy capable of doing this in digitalis. But sufficient doses must be given, whether of the tincture, powder, or infusion. The infusion, freshly prepared, is the most reliable preparation, although much of its efficiency is due to the fact that it is given in larger doses than the other preparations. Thus it is not uncommon to give f℥ss. of the infusion, which is equivalent to nearly four (3¾) grains of the powder, and thirty minims, or sixty drops, of the tincture. Yet rarely are such doses of the latter given. Less than fifteen drops of the tincture—equivalent to about one grain of the powder—is too small a dose for an adult. Fifteen drops may be given every three hours if the patient is seen daily, or oftener if the case is seen twice a day. Under such doses, if the cardiac disease is not too advanced, the urine may increase, become clear, its albumen and casts decline, and with these also the dropsy, dyspnœa, and restless, sleepless nights.

Due attention must also be paid to the bowels, for the sake of securing prompt action of the diuretics as well as the elimination which their free action secures. Even the hydragogue cathartics, such as elaterium and the salines, may be used with advantage.

The appended case illustrates so fully the cause, symptoms, effects of treatment, and morbid anatomy of this condition that I cannot but think it will form an appropriate and useful conclusion to this section:

Mary K., 39 years of age, married, was admitted to my wards in the Philadelphia Hospital on November 15th, 1873. Both of her parents suffered from rheumatism and subsequent cardiac disease, and her mother died with dropsy, which came on after confinement. The patient herself had several attacks of rheumatism, but in other respects enjoyed fair health until about five months before admission, when she began to notice shortness of breath on exertion, together with œdema of the legs and ascites. At that time she was five months advanced in pregnancy. These symptoms grew slowly worse up to the time of her confinement, after which they continued to increase even more rapidly. Her labor was quite normal, and took place five weeks before her admission to the hospital. At admission her abdomen was enormously distended, the lower three-fourths were absolutely flat on percussion, and a marked sense of fluctuation was communicated to the fingers on palpation; the lower portion of the abdomen rested upon the thighs, and the friction of the opposed surfaces had caused extensive excoriation. The legs, from the feet upwards, were œdematous, and there was also some œdema of the hands and arms. The dyspnœa was so great as to render it almost impossible for her to lie down in bed; the respirations, when she was sitting up, numbered sixty per minute.

There was no evidence of any disease of the lungs or pleura. The area of cardiac dulness was increased, and a double mitral murmur, with an aortic systolic murmur, could be distinctly heard. The pulse was feeble and frequent, beating one hundred and fifteen times a minute. The urine was found to be diminished in quantity, high-colored and acid in reaction, and to contain a large quantity of albumen, the precipitate by acid and heat being equal to at least three-fourths of the bulk of the fluid tested. The microscope revealed numerous *hyaline* and *fatty* casts. Her general health was much reduced, desire for food was almost absent, and her bowels were constipated.

She was ordered half an ounce of Basham's mixture three times daily, twenty drops of tincture of digitalis, repeated according to its action on the pulse, small doses of elaterium to relieve the constipation, and a nutritious diet, with a small quantity of whiskey (f℥ij in twenty-four hours). Under this treatment the abdomen became somewhat less tense, a larger quantity of urine was passed, and the number of respiratory movements fell to thirty per minute, but the pulse remained frequent, and seemed to be unaffected by the digitalis, although its administration was pushed to the limit of safety.

This improvement was but temporary, and on November 20th it was decided to perform paracentesis. Six pints of a highly albuminous fluid were allowed to flow from the abdomen, when the canula was withdrawn and the aperture allowed to remain open. Two hours after the operation, the pulse, which had been beating one hundred and twenty per minute, fell to fifty, then to forty-six, and became dicrotic. The dicrotism was evidently cardiac in its origin, being produced by a second effort at systole on the part of the left ventricle. The patient was, however, in every respect better; indeed, felt quite comfortable. The effect on the heart and general condition was attributed to the digitalis, which was now, however, suspended.

November 24th, the pulse was seventy-two, somewhat irregular, though entirely free from dicrotism; the respiratory movements were reduced to twenty-six, and were much more free, while the œdema was greatly diminished. She passed, on an average, three pints of urine each day. At this date the latter was found to contain an unusual number of mucous corpuscles, some spherical epithelium from the bladder, and a *few hyaline casts;* the amount of albumen was reduced to one-sixth of the bulk of the urine tested. Since the operation a large quantity of fluid, estimated at six pints, had trickled from the orifice in the abdomen. Quinine was added to the treatment.

Notwithstanding the continued use of diuretics (bitartrate of potash and infusion of juniper, infusion of scoparius, and free doses of digitalis), the ascites slowly increased, and on December 12th she was again tapped by Dr. Bruen, the resident physician, and one hundred and thirty ounces of fluid were removed. This was again followed by amendment, the œdema continuing to diminish. By December 22d the opening of paracentesis was closed, the *urine was free from albumen, and no casts were found after prolonged and careful examination.*

On December 24th, after exposure to cold, she became much worse, and by December 25th the legs were again swollen almost to bursting, and had assumed an erysipelatous hue. The respirations were hurried, the skin was hot and dry, and the urine was so charged with albumen that it became almost solid on the application of heat and nitric acid. Paracentesis was performed for the third time, fifty ounces of fluid being removed. No relief was experienced from this operation; in fact, the dyspnœa steadily increased, and the pulse became so feeble that it could scarcely be counted at the wrist.

At 3 P.M. on December 27th the pulse at the wrist was extinct and no cardiac impulse could be felt, but on auscultation the heart was found to be beating one hundred and thirty-nine times a minute. Upon extending the examination to the other parts of the chest, the physical signs of pleuritic effusion were detected on the right side. At 6 P.M., as a last resort, this side of the chest was tapped, and twenty-eight ounces of liquid removed; immediately afterwards the patient felt easier, but soon began to sink, and died at 9.30 P.M.

The general plan of treatment remained the same as that already mentioned throughout the course of the illness, although, as the indications varied from time to time, the quantity of digitalis used, as well as the form of its administration, was changed; as stated, other diuretic preparations, such as acetate of potash, juniper, sweet spirit of nitre, etc., were also tried.

At the *autopsy*, which was made twenty-four hours after death, a quantity of fluid, eighteen ounces by measurement, was found in the right pleural cavity, and the upper part of the parietal and visceral portions of the pleura was covered by a moderately thick layer of fresh lymph. The right lung was congested and somewhat compressed, but not at all inflamed. There was nothing abnormal in the left lung or pleura. The *heart* weighed fifteen and a half ounces. The left ventricle was dilated and hypertrophied. The leaflets of the aortic valve were thickened; this alteration was most marked along the edges of the leaflets and in the corpora aurantii; the latter were increased in size, being about as large as a split pea. Both leaflets of the mitral valve were thickened, and they were fused along their edges so as to form a funnel-shaped projection into the cavity of the left ventricle, with only a slitlike opening, through which the end of the thumb could scarcely be passed; the valve thus coalesced was the seat of calcareous deposit. The right ventricle was also moderately dilated and hypertrophied; the pulmonary and tricuspid valves were healthy.

The *liver* weighed ninety-seven ounces, and was very fatty. The *kidneys* were lobulated, and weighed, together, fourteen ounces; they were, therefore, somewhat enlarged, and also congested; but in other respects they appeared normal to the naked eye. On microscopical examination, the renal tubules were in part lined with healthy epithelium, or with epithelium which was only slightly granular; in other portions, however, the epithelium was highly fatty. There was also slight increase in the interstitial fibrillar tissue. No trace of amyloid degeneration could be detected by the iodine test.

DIABETES.

SECTION I.

DIABETES MELLITUS—GLYCOSURIA.

NOTWITHSTANDING the recent numerous and important contributions to our knowledge of glycogenesis, and the acknowledgment that saccharine urine, like albuminuria, is a symptom of disease rather than a disease itself, we are still far from that precise information upon which alone we dare base a classification. Until such knowledge is attained, we may define *diabetes mellitus* as distinguished from *diabetes insipidus*, as an excessive secretion of urine surcharged with sugar, associated at times with evident lesions of the nervous system, at others with as yet imperfectly understood derangements of the digestive apparatus, and especially the liver.

Residence, Age, and Sex of Patients.

Diabetes is not a very common disease anywhere, but there would appear to be differences in the frequency of its occurrence in different countries. Thus it would seem to be less common in the United States than in England. Statistics are proverbially unreliable in this country, but in Philadelphia, where we may expect to find a fair combination of the various causes of the disease, together with as much accuracy as can be expected, during the past eleven years, from 1870 to 1880 inclusive, there were 206 deaths from diabetes out of a total of 181,879, stillborn excluded, or 1 in 882, as will be seen from the following table, compiled from the records of the Philadelphia Board of Health, through the kindness of its president, Dr. William H. Ford.

Total deaths (stillborn excluded).	Years.	Deaths from diabetes.	Males.	Females.	Per cent.
15,317	1870	18	9	9	0.12
15,485	1871	17	11	6	0.11
18,987	1872	13	5	8	0.07
15,224	1873	11	10	1	0.09
15,258	1874	11	7	4	0.07
17,805	1875	20	14	6	0.11
18,802	1876	19	9	10	0.10
16,001	1877	17	10	7	0.11
15,743	1878	21	16	5	0.13
15,743	1879	26	14	12	0.17
17,711	1880	30	19	11	0.17
Total, 181,879		206	124	82	0.11

It will be seen, also, that the ratio of deaths from this cause to the total of each year is tolerably constant, and that for each year approximates that for the total.

In New York city, according to G. M. Smith, out of 80,016 deaths in three and a quarter years, 58 were diabetics, or 1 in 1379, a smaller ratio even than in Philadelphia; but the ratio in the latter city, covering a longer period, is probably more nearly correct for the entire country. On the other hand, in England and Wales, according to Dickinson, during ten years there occurred 1 death from diabetes to 632 from all other causes, and in Scotland 1 to 916. Furthermore, according to Dickinson, the disease is more prevalent in the agricultural counties of England, and of these the cooler ones—Norfolk, Suffolk, Berkshire, and Huntingdon. According to Senator, it is prevalent in France, in Normandy; it is particularly rare, statistically, in Holland, Russia, Brazil, and the West Indies, while it is common in certain regions of India, especially Ceylon, and relatively very frequent in modern times in Thuringia and Wurtemberg. Seegen says it is more frequently observed among Jews than Christians.

Diabetes mellitus most frequently affects adults in middle life, while it does also occur in children. Thus Dr. Dickinson had a case which proved fatal after six months' illness at 6 years of age; Dr. Bence Jones's youngest patient was 3½, Dr. Roberts's 3; but from the latter's work I learn that in the reports of the Registrar-General of England for the years 1851–60, ten deaths under the age of 1 year and thirty-two under the age of 3 are registered. The youngest patient I ever had was a boy of 12, who passed from under my observation before the case terminated. The disease is most frequent between the ages of 30 and 66, the oldest patient of whom I have notes being the latter age. He probably acquired the disease, however, at 64. Senator had a case in which the earliest symptoms were observed in the 69th year, and Dickinson reports one developing in the 71st year. Most important in this connection is the fact that the disease is very fatal in young subjects, recovery being almost unknown. Diabetes insipidus, on the other hand, is said to be quite common in infancy.

Very singular and inexplicable is the fact that the disease is very much more frequent in males than females, in the proportion of nearly two to one. I never happen to have had a case in a female, either in hospital or private practice. Senator's* statistics show that under the age of 20, more females are affected than males.

It so happens that of seven cases of diabetes which have recently come to my notice four are physicians, of whom two at least have been engaged in very laborious country practice. In connection with the well-known fact that anxiety, mental strain, and fatigue are admitted causes of diabetes, it is not unlikely that the peculiarly wearing life of the country physician should make him rather more than usually liable to the disease.

Pathology and Pathogenesis.

The etiology of diabetes is so intimately blended with its pathology, that it is scarcely possible to separate their consideration. What is known, therefore, of its immediate causation will be developed in connection with the pathology, while its more remote causes will be briefly considered in the ensuing section. Inseparably connected, also, with the pathology of diabetes, are the phenomena of sugar formation in the economy.

* See Senator's article on Diabetes Mellitus in Ziemssen's Cyclopædia of Medicine, vol. xvi, p. 866, ad fin.

A brief exposition of the latter seems, therefore, essential. It is very well known that during life there is constantly being produced and stored in the liver of man and the lower animals an *amyloid* substance, which was named by its discoverer, Claude Bernard, *glycogen*.[*] Its formula is $C_6H_{10}O_5$, that of starch, and the term *zoamylin* or animal starch was at one time suggested for it. The glycogen formation takes place whether animal or vegetable food is taken, but it is much more abundant upon a vegetable diet. It is commonly held that it does not occur at all with a diet of pure fats, but Salomon[†] claims that it is produced in the livers of rabbits fed on olive oil. All physiologists agree that the amyloid substance is derived *mainly* from the starchy and saccharine principles of food, but *partly* also by a splitting up and rearrangement of the elements of nitrogenous food, resulting in urea,—a soluble diffusible substance which passes into the blood,—and *glycogen*, which is stored in the liver-cells.

The most important property of glycogen is its ready convertibility at the temperature of the body into *glucose* or grape-sugar. For this is required, also, the action of an amylolytic ferment, the blood itself being such a ferment, although a separate ferment, derived either from the blood or the liver-cells, is generally considered as the agent of conversion. Bernard isolated such a ferment from the liver, and assumed that it was contained in only a certain number of the liver-cells, while glycogen was found in the others; nervous influence regulating the action of the two on each other. Pavy, on the other hand, believes that the ferment resides in the blood.

According to Bernard and his school, the conversion of starch into glucose is constantly taking place *during life*, and there is as constantly being passed into the blood of the hepatic veins grape-sugar, which is carried through the heart and lungs, and finally oxidized in the capillaries; it being held by them that there is less sugar in venous than in arterial blood. The mean difference is put at 0.3 part per 1000; the figures showing the

* Bernard, Nov. Fonc. du Foie, Paris, 1853.
† Virchow's Archiv, Bd. 61, Heft. 3, 1874, 18.

least difference being 1.10 per 1000 for arterial, and 1.08 for venous blood, those showing the greatest difference being 1.51 per 1000 for arterial, and .95 for venous blood. Formerly it was thought that the destruction takes place wholly in the lungs, but at the present day this view is, I believe, held by none. It is in the peripheral capillaries generally, and more especially in those of the muscles, that the sugar is believed to be destroyed during their contraction, the oxidation generating a part of the force produced in muscular contraction.

Dr. Pavy and his school, on the other hand, contend that to a very slight degree only does this conversion of glycogen into sugar take place in the liver *during life*. They claim that the latter organ is an assimilating one for starchy and saccharine substances, which being reduced by the action of the pancreatic juice into glucose, are absorbed as such and carried by the portal vein through the liver, whose cells pick out the glucose and convert it into amyloid substance.* According to this view there is a very slight amount of sugar in the blood, which Pavy's experiments show to be about the same in venous and arterial blood,—0.94 of one part in 1000. Corresponding to this, also, there is always in health a small quantity of sugar in the urine, 0.5 part per 1000, too small to be detected by ordinary tests, and therefore of no clinical importance. It is, however, easily recognizable by special chemical methods, as was long ago shown by Brücke.†

---

* The further steps of the assimilative action referred to, Pavy admits are not precisely known, but he considers there is every reason to believe that the amyloid substance is converted into fat, of which it is a preliminary stage intermediate between sugar and fat.[1]

† Dr. Pavy's results have been confirmed by Schiff, Henzen, Meissner, Jaeger, Pflüger, Ritter, and McDonnell of Dublin. Professor A. Flint, Jr. (New York Med. Jour., Jan. 1869), of New York, found sugar in the blood of the hepatic vein of a dog within a minute after death, by ligature of the vena cava inferior. Professor Lusk (New York Med. Jour., July, 1870) found that the blood in the right side of the heart removed during life by catheterization through the jugular vein, contained from two to four times more

---

[1] Pavy, Some Points connected with Diabetes, London, 1878, p. 110; also Pavy on Food and Dietetics, Philadelphia, 1874, p. 134; or the author's article on Food and Drink, in Buck's Hygiene, New York, 1879, vol. i, p. 165.

In diabetes, according to Pavy,[*] this assimilating action of the liver does not take place, but the glucose derived from absorption passes directly through the organ into the general circulation, and appears in the urine in quantities appreciable by the ordinary tests. Pavy says it may be that the conversion into amyloid substance occurs, and that this, through the condition of the blood, is brought back again into sugar. His own language is as follows: "The fact stands, that the sugar from ingestion is not stopped from reaching the general circulation as it ought to be, and I incline to the opinion that a simple passage through the liver occurs."[†] But Pavy admits also that in the diabetic, sugar reaches the general circulation partly by a true glycogenic action upon nitrogenous matter; since if lean meat be consumed, sugar continues to be found in the urine, although in greatly diminished quantity.

Bernard, on the other hand, held, that in diabetes the amyloid substance or glycogen is too rapidly converted into glucose to be consumed in the ordinary processes of oxidation, and therefore appears in the urine, so that in all instances of diabetes the excess of sugar in the blood is derived from glycogen, and it is an exaggeration of sugar formation rather than a lessening of its destruction.

These, then, are the two prevailing theories, according to one of which, that of Bernard, the excess of grape-sugar in the blood, which is the necessary antecedent to its presence in

---

sugar than the blood of the jugular vein in the same animals. Professor Dalton (Physiology, 1871, p. 192), by means of a comminuting instrument which enabled him to treat large quantities of liver substance in a very short time, found 1.8 part of glucose per 1000 of liver at the end of 5 seconds after death, 6.8 parts in 15 minutes, and 10.2 parts in 1 hour. Harley (Proceedings of the Royal Soc., vol. ix, p. 300) found sugar in the liver within 20 seconds after the death of the animal. The conclusion of all these last-named experimenters is that sugar *is* constantly produced from glycogen in the liver during life, but it is so rapidly removed by the circulation that only very small amounts accumulate in the organ. Hence, after death, the quantity of sugar in the liver constantly increases.

* Pavy, Some Points connected with Diabetes, London, 1878, p. 3 and p. 110.

† Pavy, op. citat., p. 110.

the urine, is due to an excessive conversion of glycogen; while according to the second, that of Pavy, it is in the main the result of a defective conversion of grape-sugar into glycogen. We are not compelled to adopt either of these views to the exclusion of the other, nor to the exclusion of any view which is consistent with experiment or clinical observations. In point of fact it will be found that we are compelled to admit both of these, and perhaps several additional explanations of the presence of sugar in the urine.

In the meantime, however, we must study also the influence of the nervous system in the production of saccharine urine. Thus, Bernard early discovered that by puncturing, by which is meant *irritating*, the medulla oblongata in the floor of the fourth ventricle, anywhere between a point 4 or 5 mm. above the nib of the *calamus scriptorius* and another about 4 mm. higher up,* the urine in a day or two acquired a considerable amount of sugar, and was increased in quantity. The amount of sugar is larger the better fed the animal, while if all food is removed the sugar is trifling in amount, or disappears altogether from the urine. The point thus punctured corresponds with the roots of the pneumogastric nerves. Hence it was supposed that the diabetes was the *direct* result of irritation of these nerves. But it was found on section of the vagus, that galvanization of the *distal* end of the cut nerve produced no effect, while the same irritation applied to the *central* end resulted in glycosuria. Whence it was learned that the pneumogastric is not an excitor but a sensory nerve of diabetes, and that the glycosuria produced by irritation of the roots of the vagi or their central cut ends, is the result of a *reflex* action called into play through the pneumogastric as a sensory, and some other nerve as a motor nerve.

Schiff† produced slight glycosuria by vertical section of the optic thalami and the great crura cerebri; and more marked

---

* The area thus bounded, which was marked out by Eckhard as the "diabetic area," corresponds very closely with the "vasomotor" area as defined by Owsjannikow. Ludwig's Arbeiten, 1871, p. 21.

† Schiff, Untersuchung über Zuckerbildung in der Leber, 1859.

glycosuria by destructive lesions of the pons and the middle and posterior crura of the cerebellum; by complete division of the cord at the level of the second dorsal vertebra, and sometimes, but not invariably, by division of the separate columns of the cord, sometimes the anterior and sometimes the posterior. Finally, Schiff produced glycosuria by section of the nerve-trunks of the limbs, as the sciatic.

Transverse section of the medulla oblongata always causes glycosuria, and as constantly section of the cord above the second dorsal vertebra, that is above the lower end of its cervical enlargement. But below this point, this effect is not produced, at least with any degree of constancy, while section of the filaments of the sympathetic accompanying the vertebral artery are again attended by it. Now these nerves are vasomotor nerves, and section of their trunks paralyzes their action and results in a dilatation of the bloodvessel walls —those of the hepatic artery chiefly—and a more rapid movement of the blood through them. Such dilatation and rapid movement of the blood is always attended by glycosuria.

Pavy* cut through the crura cerebri, completely separating the cerebrum from the parts below, without exciting glycosuria, but produced marked glycosuria by section of the sympathetic filaments ascending from the superior thoracic ganglion to accompany the vertebral artery in its canal in the foramina of the transverse processes of the cervical vertebræ; also by removal or injury of the superior cervical ganglion; also occasionally, but not always, by division of the ganglionated cord of the sympathetic in the chest. Division of *all* the nerves immediately belonging to the liver as they passed to the organ in company with the hepatic artery, hepatic duct, and portal vein, failed in every instance to occasion glycosuria. His results were communicated to the Royal Society in 1859, and published in *Guy's Hospital Reports* for the same year.

---

* Pavy, On Diabetes, 2d edition, 1869. p. 164 *et seq.* This difference in the results of Schiff and Pavy can be explained on the ground that Pavy's section was without irritation, and that Schiff's involved some irritation.

Cyon and Aladoff* produced glycosuria in dogs after section or careful extirpation of the last cervical or upper thoracic ganglion, as well as by section of the two vertebral branches, or the two nerve-filaments which form the *annulus of Vieussens* as they pass round the subclavian artery in proceeding from the upper thoracic ganglion to the lower cervical. Extirpation of the lowest cervical ganglion *after* division of the dorsal ganglionated cord of the sympathetic between the 10th and 12th ribs, or of the splanchnics, was not followed by glycosuria. But if the glycosuria had already been induced by the previous extirpation of the ganglion, subsequent section of the ganglionated cord and of the splanchnics did not cause the glycosuria to disappear.

From the result of these experiments it is inferred that the glycosuric influence leaves the cord by the filaments (*c*, Fig. 34) of the sympathetic nerve which accompany the vertebral artery, and thence through them into the lower cervical ganglion, thence by the fibres (*d*) forming the annulus of Vieussens, to the first dorsal ganglion (*e*), and thence through the prevertebral cord of the sympathetic (*f*), the splanchnics (*h*), to the cœliac ganglion (*i*), and along the hepatic bloodvessels to the liver itself, as shown in the appended diagram from Dr. Brunton's lecture on diabetes mellitus.†

It has been stated that Schiff‡ has found that diabetes sometimes results from section of the *anterior* column of the spinal cord between the medulla and fourth cervical vertebra, and as the experiments of Eckhard§ show that diabetes is not the invariable result of section of the fibres which accompany the vertebral artery, the last cervical or first dorsal ganglion, or of the fibres of the annulus of Vieussens, Dr. Brunton suggests that the glycosuric influence does not always pass from the

---

* Cyon and Aladoff, reprint from Mélanges Biologiques, and Bulletin de l'Académie Impériale de Petersbourg, vol. iii, p. 91; cited by Dr. Brunton in the paper named below; also British Medical Journal, December 23d, 1871, p. 732.

† T. Lauder Brunton, Lectures on the Pathology and Treatment of Diabetes Mellitus, reprinted from the British Medical Journal. London, 1874.

‡ Schiff, Untersuchung über Zuckerbildung in der Leber, 1859, p. 103.

§ Eckhard, Beiträge zur Anatomie und Physiologie, vol. vii, 1, 1873, p. 19.

spinal cord by the filaments above mentioned, but may *sometimes* pass further down the spinal cord and leave it by the communicating branches going to some of the dorsal ganglia,

FIG. 34.   FIG 35.

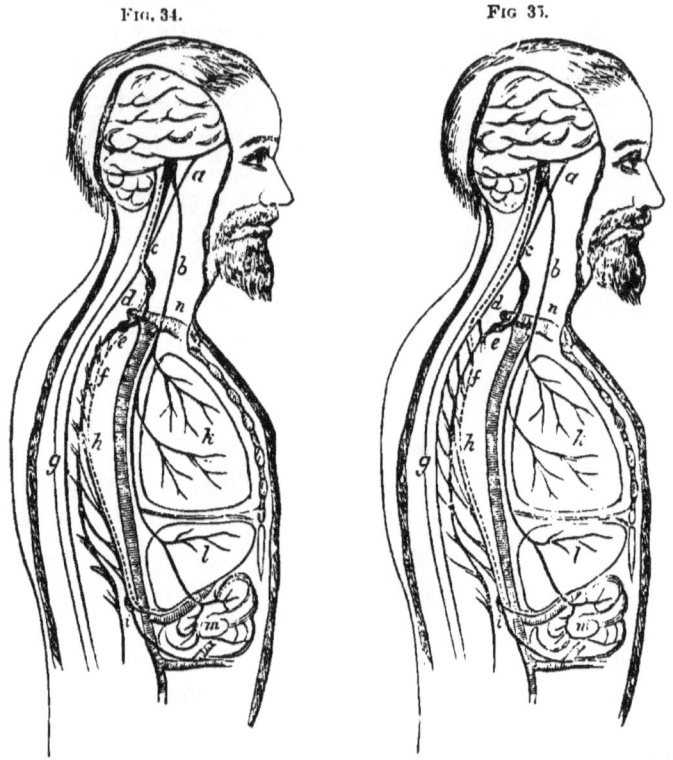

FIG. 34.—Diagram showing the course of the vasomotor nerves of the liver according to Cyon and Aladoff. These nerves are indicated by the dotted line which accompanies them ; *a*, vasomotor centre ; *b*, trunk of the vagus ; *c*, passage of the hepatic vasomotor nerves from the cord along the vertebral artery ; *d*, fibres going on each side of the subclavian artery, and forming the annulus of Vieussens ; *e*, first dorsal ganglion ; *f*, ganglionated cord of the sympathetic ; *g*, the spinal cord ; *h*, splanchnic nerves ; *i*, cœliac ganglion, from which vasomotor fibres pass to the hepatic and intestinal vessels ; *k*, the lungs, to which the fibres of the vagus are seen to be distributed ; *l*, the liver ; *m*, the intestines ; *n*, the arch of the aorta.

FIG. 35.—Diagram showing another course which the vasomotor nerves of the liver may possibly take. The letters indicate the same parts as in Fig. 34. The hepatic vasomotor nerves are here represented as passing lower down the cord than in Fig. 34, and leaving it by the communicating branches to the second dorsal ganglion ; or it may leave it, at other times, by branches to the first, or even a lower dorsal ganglion. In the latter event irritation of the third or other of the cervical ganglia may cause diabetes by being conveyed along the vertebral artery and up the cord, as indicated by the dark line, to the vasomotor centre, where it may cause reflex inhibition, in the same way as any irritation to the vagus.—After BRUNTON.

as indicated by the dotted line in Fig. 35. In this event, irritation of the lower (3d) cervical ganglion, might cause glycosuria by being conveyed along the vertebral artery and up the cord, as indicated by the dark line, to the vasomotor centre, where it may cause *reflex inhibition* as does irritation of the vagus.

Whatever may be the precise course pursued, it is evident that it is through the vasomotor fibres of the sympathetic that the glycogenic influence is regulated, whence the oneness of the centre for the two regulations. The effect of section of these nerves is easily explained, since dilatation of the vessels and increased rapidity of the movement of the blood is the consequence. And although the dilating influence may be peculiarly exerted upon the *hepatic artery* and its branches, yet increased rate of movement of the blood in the latter must also accelerate the movement of the blood in the *portal vein* and its branches. The effect of irritation upon the glycogenic function, whether propagated through the pneumogastric or some other sensory nerve, is nearly as simple. The effect of the irritation conveyed to the glycogenic centre is to *inhibit* the usual tonic influence of the vasomotor nerves on the vessel-walls. The inhibition of this influence results in a dilatation of the vessels of the liver and a speeding of their contents. Thus the same result is brought about in two different ways,—the one being direct, the other reflex.*

---

* It should be stated that the effect of irritation of the vasomotor centre is not ascribed by all experimenters to the withdrawal of an accustomed nervous influence, that is, to a paralysis, but rather to an excitation. Eckhard (Beiträge, iv, 1, 1869; vii, 1, 1873) is one of these, and he bases his objection to the usual view on the ground that, while mechanical injury to the first thoracic ganglion will produce glycosuria, no such effect follows careful removal of the ganglion, or the severing of its connections with the remainder of the thoracic chain. Bernard also, in his lectures at the College of France, in 1873, reversing his original view, concluded that the glycosuric nerve-filaments, which, starting from the floor of the fourth ventricle, he believed to pass through the substance of the cord to the fourth dorsal vertebra, and thence through the sympathetic to the liver, are *excitor* in their character and analogous to the *chorda tympani* nerve. Their irritation, according to him, produced active hyperæmia and increased sugar-formation, as irritation of the chorda tympani causes hyperæmia of the submaxillary gland and increased secretion of saliva.

Among the experimental irritations already mentioned, besides puncture of the fourth ventricle, which produce glycosuria by a reflex action, are injuries of the cerebral lobes and cerebellum, superior cervical ganglion, optic thalami, cerebral peduncles, pons varolii, middle cerebellar peduncles, cervical, sympathetic, and even sciatic nerve, and brachial plexus. It may be reasonably inferred, therefore, that pathological irritations of these same situations will result in diabetes; that if permanent the diabetes resulting will be permanent, and if temporary the diabetes will be temporary. And in point of fact, observation has in numerous instances confirmed this supposition, as will be seen later, when the pathological anatomy of diabetes is discussed.

But it is not to irritations alone applied to nerve-centres or to trunks in their continuity that glycosuria is due. Irrita-

More recently M. Lafont (Medical News and Abstract, December, 1880, p. 730, from Lancet of October 23d, 1880), isolated in dogs a segment of the spinal cord containing the first and second dorsal pairs of nerves. He next exposed the liver and irritated by faradization the isolated segment of the cord, whereupon the liver was observed to present a disseminated congestion, accompanied by a corresponding injection of the intestine and turgidity of the spleen. He also ascertained that the stimulation of these nerve-roots by a weak faradic current always caused a fall in the arterial pressure in the abdominal organs, if the root is carefully isolated. The least extension of the current to the spinal cord itself caused an increase in pressure. He found, too, that intermittent stimulation by a weak faradic current of the central extremities of the depressor branch of the vagus in the rabbit produced glycosuria; that stimulation of the central extremities of the pneumogastrics themselves produced the same result; and that division of the first two pairs of dorsal nerves caused a decided diminution of the glycosuria excited by the diabetic puncture. He found also that feeble stimulation of the central ends of the pneumogastric and any degree of stimulation of the depressor nerve always produced a fall of blood-pressure, while after division of the first three or four dorsal nerves, such stimulation produced a rise in blood-pressure.

Finally he found that painful stimulation of any mixed nerve, frequently repeated, caused glycosuria.

M. Lafont concludes from these experiments that he has confirmed the view of Bernard, that the glycosuria which results from stimulation of the central extremities of the pneumogastric in the dog, the depressor nerves in the rabbit, and the sensory nerves generally, is the result of an impression conveyed to the bulbar vaso-dilator centre, the path from which is by the cervical cord, the first dorsal nerve-roots, and the sympathetic and splanchnic nerves; and that it is irritative rather than paralytic.

tion of the peripheral distribution of nerves has a similar effect. Thus, embarrassed respiration, whether due to disease of the respiratory passages, strangulation, or inhalation of irrespirable gases, as carbonic acid, carbonic oxide, etc., and anæsthetics, as ether, chloroform, nitrite of amyl, is capable of producing glycosuria in dogs and rabbits, although the symptom has rarely been shown to attend these conditions in the human subject. In all these instances it is probably irritation of the terminal filaments of the pneumogastric—the sensory nerve of glycosuria—which causes the diabetes, by inhibiting the usual tonic action of the vasomotor nerves and producing hyperæmia of the liver. So, also, substances introduced into the blood, as woorara, and even strychnia, morphia, and phosphoric acid, produce diabetes probably in the same way, by irritating the terminal filaments of the vagi. Schiff asserts that woorara and strychnia act, not by directly irritating the filaments of the pneumogastric, but through the resulting embarrassed respiration, in the course of which the terminal filaments of the pneumogastric are irritated ; and this may be true in part. Bernard found that woorara produced, first, increased saccharinity of the blood, and, second, glycosuria when given in quantities insufficient to paralyze the muscles of respiration, and, therefore, independently of embarrassed respiration ; in other words, that it acts like puncture of the floor of the fourth ventricle, paralyzing the vasomotor nerves of the liver by inhibition. He considered that morphia acted in the same way. While admitting that this is true with regard to woorara, Lauder Brunton believes that the glycosuria which appears after injecting woorara into the blood is also in part due to the deficient consumption of sugar in the muscles which it paralyzes, and in part also to the artificial respiration employed to keep the animal alive.

Finally, Schiff[*] has shown that irritation of the liver by needles and by galvanic currents[†] passed through them pro-

---

* Schiff, Untersuchungen über Zuckerbildung in der Leber, 1859, p. 106.
† Pavy, On Diabetes, 1869, p. 137.

duces glycosuria. This is also due to resulting hyperæmia, whether direct or indirect; whence it may be inferred that congestion of the liver from any cause may produce glycosuria, although it is not unreasonable to suppose that the results of active and passive hyperæmia may be somewhat different in this respect. The hyperæmia of the liver from embarrassed respiration is passive so far as results from the backing of blood, but active so far as it is a reflex phenomenon.* That an active rather than a passive hyperæmia is essential for the production of glycosuria would appear from the fact that valvular disease of the heart is almost never attended by glycosuria, while it is always accompanied by passive congestion of the liver.

Irritation of other peripheral branches of the pneumogastric, as of the stomach and intestines, are less frequently attended by glycosuria; but that their irritation may be thus attended is shown by a case mentioned by Brunton,† in which the presence of tapeworm in the intestine caused diabetes, which was immediately cured by its removal.

Dr. Pavy discovered that the injection of oxygenated blood into the portal system produced glycosuria; also, that surcharging the blood with oxygen through the medium of respiration produced the same results. The introduction of venous blood produced no such effect; whence he concluded that oxygenated blood promotes the transformation of amyloid substance into sugar, but whether it interferes with the conversion of the sugar absorbed from the alimentary canal, or whether the formation takes place and then the transformation back again, cannot be stated. Now this, he believes, is just the state into which the portal blood is thrown by the vasomotor paralysis affecting the vessels of the chylopoëtic viscera, including the

---

* Dr. Pavy suggests (Certain Points connected with Diabetes, p. 111) another explanation of the production of sugar under these circumstances of hepatic congestion. In this state he says the *hepatic cells are subjected to undue pressure,* which may lead to more or less transudation and direct admixture of their contents with the blood.

† Brunton, op. citat., p. 14.

liver, the resulting hyperæmia causing the blood to flow through the organ too rapidly to admit its dearterialization.*

But the movement of the blood through the vessels of the liver may be accelerated in other ways than by nervous influences which dilate them. Anything which raises the blood-pressure in a part of the circulation, including the liver, at the expense of another, as by ligation or compression of a large blood-vessel, has this effect. Such causes as these are found to increase the quantity of sugar in the blood, and to cause it to appear in the urine. · Not only this, but it has been shown by Cyon that dilatation of the bloodvessels, unaccompanied by accelerated movement of the blood in the liver, is not sufficient to cause diabetes. Section of the sympathetic between the tenth and twelfth ribs, and of the splanchnics, produces dilatation of the bloodvessels of the liver, but not glycosuria, even if the fourth ventricle be afterwards punctured. Cyon's explanation is as follows: As the lower sympathetic and the splanchnics contain also the vasomotor nerves of the bloodvessels of the intestine, section of these nerves not only paralyzes and dilates the bloodvessels of the liver, but also those of the entire digestive canal. The vascular capacity of the latter canal, especially in rabbits, in which it is unusually long, is very great, indeed equal to that of the entire vascular system outside of it. Section of the splanchnics and lower sympathetic is, therefore, followed by a diversion of the blood from the liver into the digestive canal, so that, although the bloodvessels of the former are also dilated, there is a lack of blood necessary to secure an accelerated movement. If, however, the celerity of circulation in the liver be first secured, by puncture of the fourth ventricle, section of the splanchnics is not followed by arrest of sugar-formation, as the celerity of the

---

* Pavy, Certain Points connected with Diabetes, p. 96 et seq. Dr. Pavy considers that carbonic oxide acts in the same manner as oxygen in producing glycosuria, either through the direct action of the gas itself, or a compound between it and hæmoglobin; the more temporary effect of the oxygen being due to the very feeble combination between oxygen and hæmoglobin, the gas being readily displaced from the corpuscle by carbonic acid, while carbonic oxide, on the other hand, is displaced with extreme difficulty, p. 103.

circulation once established is easily maintained, in spite of subsequent section of the vasomotor nerves of the intestine.

Thus we have seen that hyperæmia of the liver, however, induced, is an important condition of glycosuria. But is it possible for glycosuria to be brought about independently of such hyperæmia? Whatever may be its bearing upon our subject, it must be admitted that there are several ways in which this may occur: 1st. There may be an excessive ingestion of saccharine and amylaceous food over and above the power of the liver to convert into glycogen. It is well known that a moderate amount of cane-sugar may be introduced into the stomach without appearing in the urine. But if an excessive quantity of sugar is thus introduced it cannot be converted into glycogen, but, passing through the liver into the general circulation, promptly shows itself in the urine. Now it is evident that such excessive ingestion may occur in man, and, as a consequence, sugar appear in the urine, while the normal converting power of the liver remains unchanged. Such glycosuria is called by Dr. Dickinson *normal alimentary glycosuria*. It is not likely, however, that this cause would ever operate in the production of a permanent diabetes. At most there could be but a temporary glycosuria.

2d. The same thing may, possibly, result from the too rapid *conversion* of starch and sugar into glucose in the intestines, in consequence of which it is introduced into the liver too rapidly and abundantly to admit its conversion into glycogen; whence its appearance in the urine. In looking about for a cause for such rapid digestion of saccharine and starchy food in the intestine it has been suggested that an excess of pancreatic ferment poured into the intestine might occasion it; and, in confirmation of this view, Niemeyer asserts that there is always hypertrophy of the pancreas in diabetes. But it is not any more common than atrophy of the same organ, and it is much more reasonable to suppose, as Brunton suggests, that if diabetes has anything to do with either condition, it is the result of an irritation of the nerves of the organ, which produce a reflex dilatation of the hepatic bloodvessels, or, with Klebs, that it is due to some concurrent changes in the cœliac axis, which produce at once the mellituria and pancreatic disease. Whatever may be the cause of such rapid conversion, it is not impossible that a certain number of cases of diabetes easily remedied by dietetic treatment may be due to it.

3d. There may be but a normal ingestion of saccharine food, which, being absorbed as grape-sugar, may still fail to undergo conversion into glycogen in its transit through the liver, and again surcharging the blood shows itself in the urine. And although such an occurrence is usually the result of hyperæmia, yet it is conceivable that it may occur independently of this condition, as from a deficiency of ferment. Dr. Dickinson believes that a certain number of cases of diabetes met with in practice are of this kind, and he calls the glycosuria thus arising *abnormal alimentary glycosuria.**

4th. It is conceivable that there may be an increased conversion of glycogen into glucose in the liver independent of hyperæmia, as the result of an increased activity of the diastatic ferment, either from increased quantity or some other unknown peculiarity. To account for such an occurrence, various hypotheses have been suggested, none of which are sufficient to account for more than a temporary glycosuria. One of these is the increased absorption of pancreatic ferment, which is thus assigned the power of converting glycogen into grape-sugar in the liver, as well as of converting sugar and starch into grape-sugar in the intestine. To account for this the hypertrophy of the pancreas already alluded to, has been also adduced.

An interesting discovery is that of Plosz and Tiegel,† that the ferment in the blood is contained in the corpuscles alone, and when the latter are destroyed, the ferment being set free acts energetically; and Griesinger has shown that the blood-corpuscles are diminished in diabetes. Whatever, therefore, may be the truth with regard to increased quantity of ferment in the blood as a cause of diabetes, there are no facts either to sustain or disprove it. It is a mere hypothesis.

---

* Dickinson, On Diabetes, London, 1875, p. 24.
† Pflüger's Archiv, 1873, Bd. vi, s. 249, and Bd. vii, s. 391.

5th. Senator* suggests that an abnormal amount of sugar may be taken up with the chyle by the lacteals by reason of some impediment to portal absorption, and carried directly into the general circulation, instead of through the liver, and thus upon the glycæmia thus induced might ensue a glycosuria.

6th. Again, if a part of the sugar ingested is normally converted into lactic acid, as is held by some physiologists, and the unchanged remainder absorbed by the radicles of the portal vein, any interference with such conversion would leave a larger residue of sugar to be absorbed by the portal radicles, and thus introduce a larger proportion of glucose into the liver than could be converted into amyloid substance.

7th. It is quite easy to conceive that diabetes may be the result of a diminished consumption of the small quantity of sugar normally supplied to the tissues by the blood. Its possibility is further substantiated by the occasional results of treatment by muscular exercise, to which especial attention has been called by Dr. William Richardson, of London, in his little work on *Diabetes and Its Treatment,*[†] and which he so successfully applied to the treatment of his own and other cases. It dare not be inferred, however, that because muscular exercise is useful in getting rid of the extra glycogen in the blood, diabetes is the result of deficient consumption of sugar in the tissues. There are, however, some other facts which have a bearing upon the subject. In the first place, Ludwig and Genersich,[‡] and Bernard[§] have shown that the blood which issues from a contracting muscle by the vein, contains much less sugar than that which enters it by the artery. Ludwig and Scheremetjewski[||] have shown that grape-sugar, as such, is not burnt off in the body, while lactic acid and glycerin are; and Bernard that as sugar disappears from the blood it is replaced by lactic acid, while Schultzen[¶] concludes from his

---

* Senator, article, Diabetes, in Ziemssen's Cyclopædia of Practical Medicine, vol. xvi, p. 952, New York, 1877.

† London, 1871.   ‡ Ludwig's Arbeiten, 1871, s. 75.

§ Bernard, Revue Scientifique, 2d series, tome iv, p. 1022.

|| Ludwig's Arbeiten, 1869, pp. 141 and 144.

¶ Schultzen, Berliner Klinische Wochenschrift, No. 35, 1872, s. 417.

experiments that the sugar, which is normally present in the blood, is split up by the action of a ferment into lactic acid and glycerin, which undergo combustion, and thus sustain the temperature of the body. If the ferment which Schultzen considers necessary for the conversion of the sugar into lactic acid and glycerin should be absent or deficient, the sugar is not split up, is not burnt up, and, therefore, appears in the urine. This Schultzen believes to be the case with certain diabetics.

· Other reasons are adduced by Dr. Brunton[*] in favor of such a view. Among these are the results of the experiments of Dock[†] and of Weiss. Weiss[‡] discovered that a considerable amount of glycogen remains in the *muscles* of starved animals after it has completely disappeared from the liver. Dock showed that after glycogen has entirely disappeared from the stomach of the starving rabbit, puncture of the floor of the fourth ventricle does not produce glycosuria, while poisoning by woorara does. Whence Brunton infers that the presence of sugar in the urine of starved animals after woorara-poisoning is due to the fact that while the glycogen in the muscles undergoes conversion into sugar, the sugar thus produced is not transformed into lactic acid and glycerin, and does not, therefore, undergo combustion as it should. Again, Dock has shown that if the fourth ventricle of the starving rabbit is punctured and then a few injections of cane-sugar made into the stomach, no glycogen is formed in the liver, and no sugar appears in the urine. If, under the same circumstance, the animal is poisoned by woorara, instead of punctured, again no glycogen is formed, but sugar is copiously present in the urine. In the former case, in consequence of the dilatation of the hepatic vessels, the absorbed grape-sugar passes too rapidly through the liver to be converted into glycogen; but it does not appear in the urine, because it is used up by the muscles, which are here intact. In the second in-

[*] Brunton, op. citat., p. 20 et seq.
[†] Dock, Pflüger's Archiv, Bd. v, s. 571.
[‡] Weiss, Sitzungsberichte der Wiener Acad., Bd. lxiv, s. 284.

stance the woorara dilates the bloodvessels by paralyzing them, just as puncture does, and the sugar passes through the liver too rapidly to be converted into glycogen. But in this instance sugar appears in the urine, because the muscles also are paralyzed by the woorara, and therefore cannot consume it.

There can be no doubt that the above reasoning is legitimate if the premises on which it is based are correct. But according to Senator, Luchsinger has disproved that in woorara-poisoning mellituria occurs without glycogen-formation, and it must be admitted that all experiments of the kind involved are more or less to be distrusted, because of the difficulties in the way of their accuracy. Senator,* in alluding to the view which would place the origin of diabetes partly in the liver, and partly in the muscles, says that " it is chiefly supported by mere conjecture, and does not comport well with a number of recently discovered facts." Of course no one claims deficient consumption of sugar in the muscles to be the sole origin of diabetes. Zimmer, who is alluded to by Senator as the author of this double view, says "the muscles are implicated, particularly in the severer cases," and such I presume to be all that is claimed by any.

Undoubtedly the sugar serves some purpose in the economy, and is, therefore, in its proper proportion a necessary constituent of the blood. Doubtless, too, it is consumed in performing its office, whether the lattter be force production by combustion, or something else, and its cessation must result in the accumulation of sugar in the blood. However insufficient the proof therefore, I feel that we dare not deny the *possibility* of such a contribution to the cause of diabetes.

It has been shown that there is a very evident relation between experimental lesions of the nervous system, glycosuria and polyuria in animals. It should be further remarked that no other symptoms of diabetes, so far as discoverable, attend these experimental lesions. It remains to consider any relation which may exist between diabetes in man and lesions of the nervous system. Some relation undoubtedly exists. Numerous cases

---

* Op. citat., p. 948.

are reported in which injuries, not only in the vicinity of the medulla oblongata and sympathetic tract, but also of the cerebral hemispheres and cerebellum, are attended by a glycosuria which is undoubtedly a consequence of them. Moreover, many of these cases have been accompanied by all the other essential symptoms of diabetes. At the same time there are many cases of diabetes, I think I may say the majority, which, not only at the beginning, but during years of observation, fail to show any symptoms which can be ascribed to lesions of the nervous system, and if towards the close such symptoms are manifest, they cannot reasonably be ascribed to primary lesions; while the most carefully conducted autopsies have failed in many cases to discover such lesions. It is true that Dr. Dickinson* has described a series of lesions about the bloodvessels, especially of the pons varolii and medulla oblongata, to be again referred to, which he makes the primary and essential lesion of diabetes. But, apart from this fact, these lesions are of a kind we would expect to find as the result of long-continued malnutrition, W. Mueller† has found them in the vessels of non-diabetics, and has failed to find them in other diabetics, while Kuelz‡ and Drs. Frederick Taylor and Goodhart§ have also failed to confirm Dr. Dickinson's observations.

Summary.

It must be admitted that the foregoing is a somewhat chaotic collection of facts bearing upon the subject of diabetes. At the same time it is necessary, where we are still ignorant of the precise pathology of a disease, to bring together all facts bearing on it, to sift therefrom such as seem to be of major importance, and from them deduce the most reasonable conclusions. From the foregoing the following inferences may be made:

* Dickinson, op. citat., p. 33.
† W. Mueller, Beiträge zur Pathol. Anat. des Rückenmarks, Leipzig, 1871.
‡ Kuelz, Beiträge zur Path. und Ther. des Diabetes Mellitus, Marburg, 1874, s. 10.
§ Taylor and Goodhart, On the Nervous System in Diabetes, Guy's Hospital Reports, 1877.

First, the liver is constantly storing up in its cells a substance identical in its composition with vegetable starch, and readily convertible by a diastatic ferment into grape-sugar.

Second, this substance, called by Bernard *glycogen*, and by Pavy *amyloid substance*, is derived chiefly from the amylaceous and saccharine articles of food, but partly, also, from albuminous food; possibly only is it formed from fat.

Third, the blood contains at all times a small proportion of grape-sugar, which in health is mainly derived from the glycogen in the liver-cells. A very minute and probably unrecognizable trace is absorbed directly from the intestine by the lacteals. So long as the quantity of sugar present in the blood is restricted within certain limits, which are not precise, but which may be put down as from .25 to .6 per cent. in dogs, sugar does not appear in the urine, but if these amounts are exceeded glycosuria occurs.

Fourth, various injuries to the nervous system, among which puncture of the fourth ventricle, transverse section of the medulla oblongata, and section of the cervical sympathetic nerve and certain of its branches are conspicuous, produce pathological glycæmia and glycosuria, which are the result of increased sugar-formation and not diminished consumption.

Fifth, it is very generally conceded that the immediate result of these nervous lesions, and therefore the immediate cause of the glycosuria thus induced, is an active hyperæmia of the liver.

Sixth, the sources of the sugar in the blood and urine under these circumstances are probably two: 1st, the *grape-sugar absorbed from the intestine*, which is carried through the liver too rapidly to permit its conversion into glycogen; and, 2d, the *glycogen itself*, which is more rapidly converted into grape-sugar by the ferment in the blood than in health. This is likely, because it is acknowledged that glycogen is very readily and rapidly convertible into grape-sugar, while the grape-sugar requires a longer time to convert it into glycogen. Hence an acceleration of the blood-current would operate differently on the two substances.

Seventh, in diabetes there exists a similar excess of sugar in the blood, which in like manner is the result of increased sugar-formation rather than diminished consumption, because although certain results of woorara-poisoning in animals go to show that diminished sugar-consumption may cause an accumulation of sugar in the blood and glycosuria, the difficulties which surround the performance of these experiments are too great and the sources of error too numerous to permit any inference from them.

Eighth, although we can conceive the possibility of diabetes resulting from simple overingestion of saccharine food, or undue rapidity of conversion of starch and cane-sugar into grape-sugar in the intestine, and more frequently from failure of the liver to intercept and convert into glycogen the normally ingested sugar and starch, in the more serious cases there is probably superadded the too rapid conversion of glycogen into sugar. This is at first derived from saccharine alimentary principles, but sooner or later comes from the albuminous principles. In no other way can we account for the emaciation and exhaustion which invariably attend the severer forms.

Finally, while we are able to trace a certain number of cases of diabetes to direct lesions of the nervous system, most frequently injuries or morbid growths in the vicinity of the medulla oblongata and cerebellum, there are at least as many in which no such association can be shown to exist. In many of the latter class, on the other hand, there is evident primary derangement of digestion as shown by their causes, their symptoms, and the results of treatment; so that the disease in these cases may be said to consist, primarily at least, in a malassimilation dependent upon derangement of the digestive organs, among which I, of course, include the liver. This organ may or may not be involved, but it is likely that in all but the simplest and most easily curable cases, it is. In what, however, the derangement immediately consists we are compelled to admit ourselves ignorant.

Causes of Diabetes Mellitus.

Excepting the *injuries to the nervous system* which are known to cause diabetes, our knowledge of its causes is not precise. With regard to the former it is well known that *blows upon the skull*, with or without fracture, and *concussions* communicated to the brain, spinal cord, and vasomotor centres in other ways, as by falls, in which the patient lights elsewhere than upon the head, are such causes. Nor does it seem necessary that injuries thus resulting need be applied to parts adjacent to the nerve-centres; blows upon the abdomen, back, legs, thorax, kidneys, and liver have all been followed by symptoms of diabetes. Succeeding such injuries to the nerve-centres, the diabetes is not always immediate but may appear weeks after the injury. The number of cases succeeding injury are however few as compared with those resulting from other causes. Thus Griesinger has collected 225 cases, of which but 13 were traumatic, being 5.7 per cent.

Next, *disease* of the nervous system, either acute or chronic, may be attended by diabetes. Among these meningitis, tubercular as well as traumatic, epilepsy, apoplexy, and tumors of the brain, especially those in the neighborhood of the medulla oblongata are conspicuous. A large number of cases thus caused have been collected by Fritz,* Goolden,† and Fischer.‡ In some of the cases thus induced there is simply transient glycosuria, while in others *mental emotion* and *anxiety* and *mental strain* have been sufficiently associated with diabetes to make it reasonable to consider them causes of the disease, especially when combined with *bodily fatigue* and *exposure*. Fright and anger may be placed in the same category. The latter two causes alone have been assigned a position among the causes of the disease.

Next may be mentioned *hereditation*. Almost all authors narrate instances wherein this cause must be acknowledged to have been in operation.

---

* Fritz, Gazette Hebdomidaire, vol. vi, 1859.

† Goolden, London Lancet, June and July, 1854, March and May, 1862.

‡ Fischer, P., Du Diabète Consécutif aux Traumatismes, Archives Générales de Médecine, 5 serie, t. xx, 1862, pp. 257 et seq.

We have only to recall the acknowledged effects of diet on diabetes to make it reasonable to suppose that the constant use of *certain foods* may cause diabetes. Such foods would of course be the farinaceous and saccharine, and it may be this circumstance that makes diabetes more common in the agricultural counties of England and in Thuringia, while it is possible also that the comparative rarity of diabetes in the United States may be due to the fact that nowhere in the world, perhaps, is animal food so freely consumed by all classes. It is well known that in some countries the people live almost exclusively upon bread and potatoes. Among these we should expect diabetes to be more prevalent. But cases directly traceable to dietetic causes are not numerous, and may occasionally be due to the indigestion which often results from so monotonous a diet.

Dr. George Harley[*] insists that *alcoholic stimulants* are a common cause of diabetes. His belief is based chiefly upon experimental results of his own, of Bernard,[†] and of Rosenstein. Harley himself showed that if alcohol is injected into the portal vein of dogs, glycosuria promptly appears; Bernard, that if alcohol is introduced into the small intestine of dogs the same result follows; while Rosenstein[‡] showed by experiments upon human subjects, that Bavarian beer and wine diminish the elimination of urea and increase that of chloride of sodium and sugar. To these results of experiment he adduces the fact that diabetes is a much more common disease in Great Britain than on the Continent, where the quantity of alcoholic drinks consumed is much less than in the former country.

Of the older authors, Willis[§] and Prout (Prout more particularly as to cider) hold the same view, but of modern observers I know no one who agrees with Harley. Dickinson, Pavy, and Senator all say that alcoholic drinks cannot reasonably be charged with being causes of diabetes, and all the cases

---

[*] Harley, The Urine and its Derangements, Philadelphia, 1872, p. 229; also, Harley, On Diabetes, London, 1866, p. 44.

[†] Bernard, Gazette Médicale de Paris, May 10th, 1856.

[‡] Rosenstein, Virchow's Archiv, 1858, s. 461.

[§] Willis, On the Operations of Medicines in Man's Body, p. 74.

of diabetes I have seen have been in decidedly temperate persons.

It is true there are some beverages containing alcohol which are contraindicated where diabetes is present, but it is more particularly on account of the sugar they contain than on account of the alcohol. These are the sweet wines and apparently new cider. At least cases are reported in which symptoms of diabetes which had disappeared, returned after free drinking of cider. It may also be said with regard to fresh beer and ale, which contain considerable unfermented sugar, that while there is no sufficient reason to believe they have ever caused diabetes, the propriety of their use by diabetics may be questioned.

Malaria, continued fevers, gout, rheumatism, the action of cold, both externally and by the use of cold drinks, have all been charged with producing diabetes without sufficient reason. Dr. Pavy relates a case associated with extravagant sexual indulgence.

Pathological Anatomy.

Except the lesions of the nervous system occasionally causing it, diabetes can hardly be said to have an essential morbid anatomy. That is, except in the instances alluded to, the morbid lesions which are found after post-mortem examination to have been associated with diabetes are, in the main, such as are the consequence of the continued presence of the condition, rather than such as cause the symptoms. Sometimes there are absolutely no alterations discoverable either by the naked eye or the microscope.

To begin with the organ which is apparently immediately responsible for the symptoms, the *liver*, it frequently presents the appearances of a hyperæmic organ; that is, it is *darker* and *harder* than the normal organ, while it is also enlarged, sometimes extremely, at others only slightly. The maximum enlargement reported is three times the size in health. Corresponding to this, the microscope, by very moderate amplification, shows enlarged and distinct acini, with capillaries dilated and distended with blood in various degree. Higher magnifying powers—300 to 400 diameters—show the liver-cells to be enlarged, distinctly nucleated, rounded, and disposed to fuse into each other. If a weak solution of iodine is added they strike a wine-red color, which, according to Rindfleisch, is confined to the nucleus; but, according to Senator, may extend to the whole of the cell. Klebs ascribes this reaction to postmortem changes in the glycogenic substance. The minute changes described are said by Rindfleisch to be more striking in the peripheral zone of the lobule, that of the *portal vein;* while the intermediate zone, or that of the hepatic *artery*, is fatty, and the central part, including the rootlets of the hepatic *vein*, is nearly normal. Stockvis and Frerichs assert that the enlargement of the liver is partly contributed to also by a new formation of liver-cells. In one instance Dickinson* found, in addition to general venous thrombosis and apparently in connection with it, patches of a remarkable spongy transformation, which he ascribed to " extravagant dilatation of the capillaries belonging to the hepatic vein." Under the microscope sections had a worm-eaten or honeycombed look. The threads of the network were chiefly composed of glandular epithelium compressed and elongated; the cavities were empty, and proved to be contorted tubes, which opened into branches of the hepatic vein; whence he concluded them to be dilatations of the capillaries in connection with this venous trunk. The description suggests that of an angeioma of the liver.

Dickinson describes an overgrowth of connective tissue as well as of epithelium resulting in a " hypertrophic cirrhosis," to which Trousseau also alludes in connection with diabetes, and of which Budd† reports a single case. Klebs says that as the disease continues, the liver again becomes reduced, and a diminished size has occasionally been observed.

A diminution in the normal proportion of fat in the liver-cells is quite a regular occurrence, according to Beale, and quantitative analysis by Folwarczny‡ sustains this view, while Frerichs also found the hepatic cells destitute of fat in four

---

* Dickinson, op. citat., p. 50.

† Budd, Diseases of the Liver, London, 1845.

‡ Leberanalysen bei Diabetes mellitus, Wiener Zeitschr., N. F., 1859, ii, 6.

cases and greatly diminished in another. This observation is
very interesting in connection with the statement of Pavy that
glycogen is normally converted into fat, which is stored up for
further use in the economy, but that in diabetes this conversion
fails to be made.

As to the *pancreas*, the statement of Niemeyer,[*] that "*hypertrophy* of the pancreas occurs with remarkable frequency, con-
sidering how seldom this organ is the seat of disease," has already been alluded to. Senator[†] also says this organ "is found
diseased with surprising frequency, in particular either *atrophied* or, in addition, degenerated." These diverse statements
illustrate the true position of the morbid anatomy of diabetes.
It may be well to mention, however, the changes included by
Senator under atrophy and degeneration. "Sometimes," he says,
"the degeneration consists merely in primary fatty destruction
of the gland-cells, but sometimes it is induced by cancer, by
the formation of calculi, and by obstruction of the efferent ducts
with cystic dilatation of the body of the gland. In certain cases
the wasting of the gland has reached the highest degree, so that
scarcely any discernible remnant of the secreting parenchyma
was to be found." He says further : "The frequency of these affections of the pancreas was not noted until somewhat recently, in
consequence of Bouchardat's contributions. Only isolated data
in regard to this point have come down to us from earlier times,
such as the discovery of calculi in the pancreas of a diabetic by
Cowley, and that of cancer by Bright. Griesinger, who had found
the pancreas atrophied in one of the five diabetics whose bodies
he examined after death, still believed that this lesion was of
no significance whatever. But the observations which have
been published in great numbers (Hartsen, Fles, v. Reckling-
hausen, Frerichs, Klebs, Harnock, Kuelz, Schaper, and others)
allow us to assume that diseases of the pancreas are present in
about one-half of all the cases of diabetes. Among nine cases
Frerichs saw atrophy or fatty degeneration of the gland five

---

[*] Niemeyer, Textbook of Practical Medicine, translated by Hackley.
Third American edition, 1870, vol. ii, p. 751.

[†] Senator, loc. citat., p. 887.

times, and in the Vienna dead-house the pancreas was found strikingly small, soft, and anæmic in thirteen out of thirty diabetics. (Seegen.)"

To the above may be added the following cases mentioned by Dr. Richardson,[*] one of Dr. Elliotson's in which the pancreatic duct and large lateral branches were crammed with white calculi, and four successive cases of Dr. Hyde Salter, in which the pancreas was degenerated, so that he thought he had discovered the pathology of the disease.

It may be, as Senator says, that this cannot be an accidental coincidence. He believes further that Klebs's view, that the coexistence of diabetes mellitus and diseases of the pancreas depends upon *lesions of the cœliac plexus*, is the best founded: "Either the disease (cancer, formation of calculus, and inflammation of the surrounding tissue) starts from the pancreas, encroaches upon the plexus, and gives rise to diabetes by destroying its ganglion, or else the cœliac axis is first affected, and in consequence thereof circulatory disturbances arise in the territory supplied by the cœliac artery, which lead to degeneration and atrophy of the pancreas."

In 1877 Lancereaux[†] presented to the French Academy of Medicine specimens of profound lesions of the pancreas from patients who had died of diabetes mellitus, and argued for their causal relation to the symptoms of the disease from the effect of extirpation of the pancreas in the lower animals, which became voracious, grew thin, and died speedily.

It would appear from the above at least, that in the future closer attention should be given to this organ than in the past.

The *kidneys*, primarily unaffected, are undoubtedly sooner or later influenced by the constant hyperæmia to which they are subject in eliminating the sugar, although, as the hyperæmia is an active one and there is a free movement of the blood through the organ, it is not to be expected that this

---

* Op. citat., p. 65.
† La France Médicale, Nov., 1877.

should appear early or be invariably present. The appearances commonly met are those of hyperæmia and overgrowth of epithelium; in a word, those of catarrhal nephritis. Occasionally the changes are more advanced, and the epithelium is fatty. These changes need not necessarily be attended by albuminuria previous to death.

When, however, albuminuria is present in diabetes, as is not very unfrequently the case, the lesions of the kidney described may be expected as secondary to the primary disease; although it may also result from amyloid degeneration of the kidney, which may itself be due to the exhaustive drain to which the organism is subject, or to the phthisis which so constantly supervenes upon diabetes mellitus.

As to the proportion of cases in which the kidneys reveal morbid alterations, it is a decided majority. Thus Griesinger found them in 32 out of 64 autopsies, Seegen in 20 out of 30 cases examined at the Vienna dead-house, and Dickinson in 25 out of 27 autopsies at St. George's Hospital, London.

Catarrh of the pelvis of the kidney and of the ureters is mentioned by Senator as rather frequently found, and as due partly to the final complications which prove fatal, and partly to the irritating effect of the sugar and other abnormal elements of the urine.

Atrophy of the *testes* is mentioned by the same author, on the testimony of Romberg and Seegen, as occasionally present in young persons.

The *lungs* are almost invariably the seat of cheesy deposits and cavities resulting from their softening, the result of circumscribed catarrhal pneumonias, which so constantly attend the latter stages of diabetes. The changes in the lung are clearly not tubercular, but cheesy, nor is tuberculosis prone to occur by infection from the cheesy foci. *Gangrene* of the lung is sometimes present.

Many other isolated lesions are described as occurring in diabetes mellitus, but as they bear no necessary or evident relation to the disease they attend, they need not be especially mentioned. Among these are gastric and intestinal catarrhs, hæmorrhagic erosions of the gastric mucous membrane, pleuritic exudations, etc.

The Alterations in the Nervous Centres described by Dr. Dickinson as the Essential Morbid Anatomy of Diabetes.

In addition to the more palpable lesions of the nervous system so often alluded to, which are attended by glycosuria as an isolated and more or less harmless symptom, Dr. Dickinson claims that there is found, upon close examination even by the unaided eye, a set of changes which are essentially associated with diabetes, and which may be said indeed to constitute its pathology, insomuch that he defines *diabetes as a disease of the nervous system characterized by saccharine urine.* In consequence of the extreme importance of this subject, I extract almost entire Dr. Dickinson's description of these changes:

"They consist, to the naked eye, of a fine porosity or cribriform appearance in limited patches of the white matter, as if closely beset with pinholes, each puncture containing a vessel much smaller than itself. More rarely, considerable cavities, such as might hold peas, are seen, especially in the pons, in connection with one of the processes of the pia mater. . . .

"The microscope is not necessary for their detection, though it is for their description. For their recognition it is only necessary that the brain should be looked at while fresh, or for their more clear display should be hardened in any way which allows of the exposure of clear and hard sections. . . .

"The excavations are found about arteries, or in positions which arteries once occupied. . . . They are caused by a destruction and absorption of the nervous matter along the course of arteries, and are, at least in some instances, caused by an escape of the contents of the vessel into the surrounding tissues, with consequent degeneration, softening, and removal of the nervous matter which has been permeated by the intrusion. The escape appears to be rather of corpuscles by migration, than of blood in bulk by rupture. . . .

"When the disease has proceeded to its natural end, the excavations are widely scattered through the brain; numerous, small, and closely set in the white matter of the convolutions, fewer and larger about the central parts. The corpora striata, optic thalami, pons, medulla, and cerebellum are the chosen seats for the largest and most striking of the holes, in which situations the cavities are determined by the course of considerable arteries or by penetrating folds of the pia mater. . . .

"These holes are evidently exaggerations of the perivascular spaces. . . According to their date they contain degenerate remains of nerve-tissue, remnants of vessels or of extruded blood, or are empty. The products of nervous degeneration are first removed, then for awhile the cavity contains only dilated or shrunken and obsolete arteries, with areolar tissue, derived apparently from the perivascular sheath, and crystals of hæmatin. Lastly, these disappear also, and mere vacuity is left.

"In rapidly fatal cases, the cavities are sometimes filled with a translucent gelatinous substance, containing, besides vascular elements, the granular or globular products of nervous disintegration, with delicate fibrillæ and nuclei, derived, in part, from the perivascular sheath, and partly from the condensed remains of the connective tissue of the destroyed nervous substance.

"This transparent substitute for brain-matter is soft and elastic, and often eludes the edge of the razor, so that, although conspicuous enough to the naked eye, it requires some care to obtain sections for the microscope.

"Among the contents of such cavities it must be mentioned that large nerve-cells, displaced and somewhat degenerated, are sometimes seen among the débris, as if such cells, the place of which had been usurped by the excavations, survived the destructive process longer than the nerve-fibre.

"In the more chronic forms of the disease, as it occurs in elderly persons, the excavations are usually empty, though the remnants of nervous decay are usually to be found fringing their margins or collected as an irregular sheath upon the dilated or shrunken artery.

"The changes in the cord are similar to those in the brain, but less declared. Erosions about the arteries are evident, especially in the transverse commissure, the white band of which is sometimes completely divided in the track of one of its large vessels. Holes such as have been described in the brain, sometimes, though rarely, perforate the gray horns. The most striking change in the cord, however, is dilatation of the central canal, which in the dorsal and lumbar regions is sometimes expanded to many times its normal diameter, and forms a conspicuous object immediately the cord is divided. This expansion of the channel (sic) is not constantly present, but when it is, it is sufficiently remarkable. I am not aware, at present, how far it is peculiar to diabetes."

With regard to these views of Dr. Dickinson's I will here merely repeat what was mentioned on page 247, that W. Mueller has found these changes in the muscles of non-diabetics, and has failed to find them in other diabetics, while Kuelz and Drs. Frederick Taylor and Goodhart have also failed to confirm Dr. Dickinson's observations. Also, that others allege these alterations are secondary, being the result of long saturation of the bloodvessels and tissues with sugar, which finally comes to pervade all tissues of the diabetic. To this Dr. Dickinson replies that in such an event the seat of emigration or of rupture would be in the area of the capillaries rather than of the arterioles.

Symptoms, Course, Duration, etc.

Almost invariably the earliest symptoms noted by the diabetic are *thirst* and *frequent micturition*. One or the other of the two may be noticed first, or the patient's attention may be called to both simultaneously. It occasionally happens that a *dryness of the fauces* and a glutinous viscid character of the saliva attracts attention before any other symptom. Frequently the first words addressed to the physician are, "Doctor, I am burning up with thirst," or it is observed that a drop of urine falling upon the boots or clothing, and there evaporating, leaves a persistent white spot, which is sugar. A *dryness* and *harshness of the skin*, due to absence of perspiration, soon make their appearance and early attract the attention of those who ordinarily perspire easily, and occasion varying amounts of discomfort. *Itching of the skin* is also sometimes present. Notwithstanding the dryness of the skin, the temperature of the body is not increased, at this stage scarcely altered, although later in the disease it is decidedly lowered. If the further progress of the disease is not arrested by reason of the physician's attention being called to it, a *voracious appetite* becomes the next symptom, but, notwithstanding the latter, the patient observes that he slowly loses in weight and grows daily weaker. The rate at which these symptoms succeed each other is not uniform. Sometimes it is with great rapidity, at others the successive stages are exceedingly slow in developing.

The above category includes all the symptoms which present themselves in the milder form of the disease, such as occurs in adults past middle life, and is ordinarily quite amenable to treatment. But unless averted, all of these symptoms become intensified. The patient complains of constant burning thirst, is constantly urinating and as constantly drinking water to quench his thirst, and, while eating enormously, grows emaciated, although at the onset of the disease he may have been a robust and even portly man.

Dyspeptic symptoms appear at various stages, seldom very early, because they are generally the result of the large amounts of food ingested. Acid eructations, flatulence, and epigastric pain, or an indescribable sensation spoken of as "sinking" of the epigastrium, are among them. Constipation, probably in consequence of the free ingestion of food and abundant waste therefrom, is not, in my experience, a very early symptom; but, sooner or later, the general "dryness" makes itself felt here, and more or less obstinate constipation results. On the contrary, diarrhœa sometimes is present.

Soon succeeding the constipation is a peculiar vinous or acetous odor of the breath, which has been compared to that of stale beer, and by Sir Thomas Watson to the odor of a place in which apples are kept. This is believed to be due to *acetone*, and probably also to *alcohol*, both of which exist in the blood of profound diabetics during life, and both of which are derived from the sugar.

Cough sooner or later presents itself as the result of bronchitis and catarrhal pneumonia, and, with the copious expectoration incident to them, adds to the debilitating agencies already at work. Roberts thinks phthisis occurs in one half the cases. The consumption thus induced sometimes rapidly hastens the fatal termination, while at others it appears to have but a trifling influence in this respect. The other symptoms characteristic of pulmonary consumption are also present, not even excepting hectic sweats. The perspiration thus arising may contain sugar.

The early *loss of sexual desire* is said to be characteristic, but whether more so than in exhausting diseases generally I cannot say from my own experience.

In advanced stages the *temperature* of the body, unless influenced by intermittent febrile disease, is almost invariably lowered from $1°$ to $2\frac{1}{2}°$ F. Dr. Dickinson refers to the case of a boy of six, in whom the temperature ranged from $93.6°$ to $94.8°$ F., and who died of pneumonia, during which the temperature rose to $97.8°$ F.

The above-recorded symptoms include those which may be considered essential to the disease, or at least are invariably present if it remain unchecked. There are many others which occur more or less frequently, but not constantly, in its course.

Among these is *cataract,* the association of which with diabetes was long ago noted by Prout. Griesinger found it had been present twenty times in 225 cases collected, Bouchardat once in 38, Roberts once in 45, and Dickinson once in 28 cases dying of diabetes in St. George's Hospital, London. It usually occurs in advanced stages, but has even been the first symptom of the disease noticed. It develops rapidly, and is nearly always symmetrical, involving both eyes simultaneously, but not to the same degree.

The experiments of Kunde,[*] Dr. S. Weir Mitchell,[†] of this city, and Dr. B. W. Richardson[‡] led them to conclude that the cataract is due to an exosmotic loss of fluid from the lens, but Von Graefe, from the vantage-ground of a large experience, concluded it was the result of an impaired nutrition, itself caused by the vitiated sugar-laden blood.

In an admirably studied case of diabetes occurring in the practice of Dr. Louis Starr, in which there was also cataract, Dr. Albert G. Heyl[§] discovered and pictured an extraordinary condition of the fundus not heretofore described, to which he gives the name *intraocular lipæmia.* This condition is characterized by the light-salmon color of the blood contained in the branches of both retinal vein and artery, as contrasted with the yellow-red color of the arteries and the dark cinnabar-red of the veins in health, by the apparently large diameter of these vessels, and by the very light color of the fundus, these appearances being due to the presence of molecular fat in abnormal amount in the blood.

This appearance is reproduced in the colored frontispiece, which was lithographed from a painting of the right fundus of the patient, by Dr. C. B. Nancrede. The appended description of the plate is furnished by Dr. Heyl.

"A glance at the picture will show that there is no indication

---

[*] Kunde, Wurzburger Verhandlung., vii, 1856; Archiv für Ophthalm., Bd. iii, p. 275.

[†] Mitchell, On the Production of Cataract in Frogs by the administration of Sugar, Amer. Journ. Med. Sci., N. S., vol. xxxix, 1860, p. 106.

[‡] Journal de la Physiologie, par Brown-Séquard, 1860.

[§] Lipæmia and Fat Embolism in Diabetes Mellitus, New York Medical Record, vol. xvii, 1880, p. 477.

of intraocular inflammation, *i. e.*, no tortuosity of the vessels, obscuration of the scleral ring, hæmorrhage, structural change in retina or choroid. The marked features which arrest attention are:

"1. The peculiar color of the fundus; it may be described as a salmon-red. The painting was made with the aid of gaslight and the ordinary ophthalmoscope with concave mirror. The color of the fundus thus obtained will be best represented by taking the picture to a well-lighted window and allowing the direct rays of the sun to fall upon it.

"The retinal vessels were of nearly the same color as the rest of the fundus. When the light from the ophthalmoscope fell directly upon them the appearance of the vessels was as represented in the picture; if, however, the handle of the mirror were slightly rotated, so as to deflect the light, a dark axis could be seen in some of the vessels, which were probably veins. By color alone, arteries could not be distinguished from veins.

"2. The breadth of the retinal vessels. It was apparently about double what it ought to have been under normal circumstances. This was not caused by distension and increase in the vessel's calibre, but is to be explained in this way: When a bloodvessel of a living tissue is examined with a microscope the blood-current is seen to consist of two portions: (a) a peripheral portion, consisting of transparent serum; (b) an axial portion, containing the corpuscles. Now, under normal circumstances in ophthalmoscopic examinations, the axial current alone is easily seen, and what appears to be the breadth of a retinal vessel is really only the breadth of the axial current; owing to the transparency of the peripheral portion it can only be seen by a trained eye, and even then only under favorable circumstances. If, however, the serum of the blood become sufficiently opaque it will be visible, and the retinal vessels will appear of full width, which is about double the diameter of the axial current. This is what happens when finely divided fat exists in sufficient amount in the serum as in this case of lipæmia. It must be remembered that the abnormal appearances just described require for their production a sufficient amount of finely divided fat. It is possible, as I have experienced in a case of diabetes mellitus, for numerous oil-molecules to be visible under the microscope, and yet no appreciable change to occur in the gross appearances of the blood or in the fundus of the eye.

"The only known condition with which intraocular lipæmia is liable to be confounded is that seen in the eyes of leuco-cythæmic patients; for a discussion of the differential diagnosis I must refer to my paper in the *Trans. Amer. Ophth. Society* for 1880. As in ophthalmoscopic examination the details of the background of the eye appear magnified, the details of the picture have been represented as magnified 11.25 times. It ought also to be stated that the patient from whom the picture was taken had very light hair, irides, etc.; the fundus therefore, under normal circumstances, would have been light in color, but not to the extent represented."

Dr. Dickinson says also, that the ophthalmoscopic examinations of his diabetic cases by his colleagues, Messrs. Power and Brudenell Carter, mostly revealed dilatation of the retinal vessels. In one instance only was anæmia observed. Finally, atrophy and hæmorrhagic and inflammatory affections of the retina have been observed.

Among *functional* derangements of vision said to occur are amblyopia, presbyopia, and loss of accommodating power from defect of the ciliary muscle. Occasionally total blindness has occurred from atrophy of the retina.*

Other derangements of the special senses said to attend diabetes are impairment of hearing, roaring in the ears, and derangements of smell and taste.

Boils and *carbuncles* in the skin are also of more or less frequent occurrence as consequences of the malnutrition growing out of diabetes, although it is said that the former are occasionally the first symptoms recognized. The latter never occur early, but when present are frequently the immediate cause of death.

Gangrene of various parts of the body is another of this class of symptoms. It is sometimes spontaneous, but more frequently is immediately caused by some trifling injury, which

* Dufresne, De l'Amblyopie Diabétique, Gaz. Heb., November, 1861.

under other circumstances would be without result. It has been known to start from a blister. This mode of origin makes it unnecessary to seek any farther immediate cause such as inflammation, degeneration, obliteration of arteries, etc. Beginning most frequently in those parts of the body most remote from the centre of the circulation, as the toes, its progress and appearances are like those of senile gangrene.

A *spongy state of the gums* with recession and excavation are sometimes present, resulting in extreme cases in absorption of the alveolar processes and falling out of the teeth.

Albuminuria is a symptom which sometimes attends the advanced stages of diabetes. It is the result of the alterations in the kidney already described (p. 255), as consequent upon the long-continued hyperæmia. It is not usually large, but may be considerable.

Eczema, with itching and burning of the labia and vicinity, is a symptom sometimes met in females, which is incident to the extremely frequent micturition. In the male, the meatus urinarius is sometimes the seat of a similar irritation.

Unilateral sweating has been observed. Senator refers to three cases, one by Koch and Nitzenadel in a man of 30, where the left half of the face was involved; one by Kuelz in a man of 51, in which the left half of the face was affected, and a third by the same observer in a man of 46, in which the right half was involved. Some lesion of the sympathetic is believed to lie at the bottom of this symptom.

Œdema, which also sometimes appears late in the disease, is not usually the result of the renal involvement. This symptom, when coincident with the enormous diuresis, is a truly remarkable one. It is usually explained on the ground of the profound anæmic cachexia which is always established before it appears, but it is difficult to conceive the transudation of water from so dense a fluid as the blood under these circumstances without some resistance to its onward movement. Such resistance might be afforded by the viscidity of the sugar-laden blood. On the other hand, we can conceive the ingestion of fluid by reason of thirst to be so large that the kidneys cannot, with sufficient rapidity, eliminate it; whence a filtration of it into the tissues.

The term *diabetic coma* has been applied to a form of coma which sometimes is the immediate cause of, or at least immediately precedes, death. The condition is one of suddenly or gradually supervening unconsciousness, with or without previous irritability or uneasiness. Convulsions do not occur. In addition to coma there are frequent and feeble pulse, rapid and deep inspiration. It has been variously ascribed to poisoning by sugar, acetone, alcohol, or other unknown substances, and by Professor Sanders and D. J. Hamilton* to slow carbonic acid poisoning due to fat embolism of the pulmonary

FIG. 36.

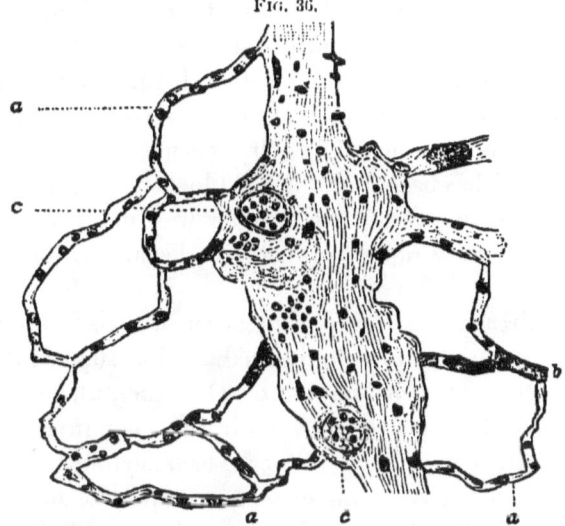

a, Supposed fatty emboli in the capillaries of the lung, stained black by perosmic acid; *b*, oblong, branching masses of the same; *c*, transverse sections of arterioles showing globules of fat among the blood-corpuscles.

vessels, the result of lipæmia. In Dr. Starr's case of diabetes, alluded to on p. 261, a careful study by Dr. J. H. C. Simes of sections of the lung, treated by perosmic acid, "demonstrated the fact that the pulmonary bloodvessels were occluded by fat emboli," as shown in Fig. 36. But as Dr. Starr correctly says, the share of the embolism in producing the coma

* Edinburgh Med. Journ., July, 1879. The conclusions of these gentlemen were based upon the clinical histories and the results of the post-mortem

and death of the patient is very uncertain, since on account of his debilitated condition a croupous pneumonia which supervened, although limited in extent, was quite sufficient to determine the fatal issue.

Alterations in the Blood, Urine, and other Secretions.

The Blood.—It has already been mentioned, that in diabetes the blood becomes highly charged with sugar, and that acetone and alcohol are also found in it. Both of these latter are derived from the sugar. From the presence of the first, we should naturally expect a viscidity and higher specific gravity of the blood serum, which has been found as high as 1033, the specific gravity of the normal serum being 1028. On the other hand, the serum has been found thinner than normal, containing, according to different analyses, from 80.2 to 84.8 of water instead of the normal 78–79 per cent. From the fact that no sugar at all has been found in the blood in certain cases after death, it does not follow that sugar was not present before death, since it very rapidly undergoes decomposition in cadaveric blood.

I have already mentioned that according to Griesinger the *red* blood-corpuscles are diminished. This statement is sustained by the study of the case of Dr. Starr, already alluded to, in which the corpuscles were carefully counted by Dr. F. P. Henry by means of a Gower's hæmacytometer with the following result, viz.: number of *red* corpuscles to cubic millimeter, 4,205,000, the normal being at least 5,000,000; number of *white* corpuscles to cubic millimeter, 50,000, or 1 white to 84 red cells, the normal being 1 white to 350 or 400 red cells.

---

examinations of several cases. For these I must refer the reader to their original paper in the Edinburgh Medical Journal. A summary of the data leading to their hypothesis will be found in Dr. Starr's paper on Lipæmia and Fat Embolism in Diabetes Mellitus, published in the New York Med. Record, vol. xvii, 1880, p. 476.

Very interesting, in connection with the pathology of diabetes, is the presence of fat in the blood, producing the technical *lipæmia*. This was observed by the earliest students of diabetes, and is attested by many analyses, as well as by the milky appearance of the serum, and the intraocular appearances described by Dr. Heyl. The analyses of Simon show from 2 to 2.4 per cent. instead of the normal 1.6 to 1.9 per cent.

The Urine.—The most noticeable peculiarity of diabetic urine, to the patient at least, is its *enormous quantity*, which has been known to exceed 70 lbs. (31.78 kilograms) in 24 hours, while apocryphal accounts of larger amounts are extant. Frank records 52 lbs. (23.6 kilograms). Bardsley* 36 pints (20.4 liters) and 32 pints (18.16 liters). Bence Jones found 56 pints (31.78 liters), Sir Thomas Watson and Dr. Dickinson 26 pints (14.77 liters), and Dr. Pavy 32 pints (18.16 liters). From 70 to 100 ounces (2100 to 3000 cubic centimeters) are common quantities.

The quantity of urine passed may be put down as limited by the amount of fluid ingested. For while it is possible that the amount of urine secreted may exceed for a very short period that ingested, it is evident that this cannot continue for any length of time, and in point of fact it is found to be almost invariably a little less, the remainder being removed by the lungs, skin, and bowels.

On the other hand, it was early observed by Cowley† (1778), that the quantity of water occasionally is not at all, or but slightly increased. To this condition Frank,‡ another old author, gave the name of *diabetes decipiens*. But all modern observers unite in saying that this phenomenon, though occasionally occurring, is but a temporary one. It may characterize the beginning, or appear in the course of the disease.

It is well known, also, that intercurrent disease, especially

* Bardsley, article on Diabetes in the Cyclopædia of Practical Medicine, Philadelphia, 1845, p. 607.

† Cowley, Th., London Medical Journal, 1788.

‡ Frank, J. P., De curandis hom. morbis epitome. Lib. v, De Profluviis, Pars. i, Manheimii, 1794.

febrile affections, sometimes diminish the quantity of urine as well as the amount of sugar excreted; while the same diminution of urine and sugar also occasionally occurs towards the fatal termination of the disease.

But the most important change in the urine is the *presence of sugar*. Of this the quantity varies greatly in different cases and at different times in the same case. Every case of trifling and temporary glycosuria should not however be considered a case of diabetes. The sugar should be easily recognizable by the ordinary tests, and should be constant. From what may be indicated as "evident traces" the proportion of sugar may reach as much as 15 per cent. The twenty-four hours' quantity varies similarly. The maximum quantity secreted in this time appears to be that reported by Dickinson, wherein a man 25 years of age voided 50 ounces, or 1500 grams, in 24 hours. But the more usual quantity is from 10 to 80 milligrams to the cubic centimeter, or from 20 to 25 grams in 24 hours, according to the metric system; this corresponds nearly to from 5 to 30 grains to the fluid ounce of the English system, or from 300 to 3800 grains in the 24 hours.

The effect of intercurrent febrile disease in producing a diminution in the amount of sugar excreted has been alluded to; also a similar decrease and even disappearance towards the fatal termination of a case. The possibility of this occurrence should be remembered, lest an illusory improvement thus produced be mistaken for an actual one.

A diminution and even disappearance of sugar from the urine has been observed by Bouchardat, and more recently by Kuelz, in consequence of muscular exercise, while it is scarcely necessary to say that accidental as well as intended changes in diet are followed by consequent variations. So, too, urine passed after fasting, as on rising in the morning, contains generally less sugar than that passed after a meal, and in testing urine whence sugar is disappearing, it is well to remember this circumstance also, as sugar may entirely have disappeared from a urine passed on rising, while it may be present in that passed after a meal.

Consistently with this increased amount of solid matter in solution, the *specific gravity* of diabetic urine is, as a rule, high, 1040 being very common, while Bouchardat found it as high as 1074 in one instance. The well-known disposition of diabetic urine to become *frothy* on shaking, and to maintain this frothy condition, is a natural physical result of its increased density. The urine of lowest specific gravity in which I have found sugar easy of detection was 1018. Pavy, however, records a specific gravity of saccharine urine as low as 1010, and Dickinson as low as 1008. Such low specific gravities may be due, as Senator suggests, to previous destruction of sugar by fermentation, as well as to low proportion of other normal ingredients.

Concurrent with the increase in quantity of urine is a *paleness*, which proceeds in extreme degrees to an almost absolute absence of color, so that the urine, at least in moderate bulk, is as colorless as spring-water. This clearness may be diminished by exposure, and almost all diabetic urine, sooner or later after exposure at a moderate temperature, becomes a little cloudy from the development of fungi coincident with fermentation.

The *odor* of the urine is usually normal when first passed, but sooner or later, in consequence of fermentation setting up, it acquires an acetous odor. The latter change also increases the degree of the normal acid reaction, and maintains it much longer after exposure to the air than is the case with normal urine. This acetous odor is ascribed to acetone and alcohol. The urine may have a sweetish odor when passed, an odor which has been compared by one of my correspondents afflicted with the disease, a very intelligent physician, to "sweet brier."

Along with the absence of color in diabetic urine, the absence of *sediment* is a conspicuous feature. This is not invariable, however, for a copious *uric acid* sediment is sometimes early present, and at others appears sooner or later coincidently with the increased acidity of the urine resulting from fermentation.

It is thought by some to be a favorable symptom. In the sediment may also be included the fungus known as *penicilium glaucum*, common to acid urine, as well as the more characteristic yeast or sugar fungus, the *torula cerevisiæ*. This also sometimes appears as a *mould* on the surface of the urine.

Of the normal chemical constituents of the urine *urea* is almost invariably increased. This is contributed to by two causes, and possibly by a third. The first is the ingestion of large amounts of nitrogenous food, whether to appease the appetite or by the physician's advice. Such ingestion is, of course, followed by an augmented excretion of urea. The second cause is the increased discharge of water by the kidneys, such increase always carrying out with it an increased quantity of urea, washed out of the tissues, as it were, by the water passing through them. The third cause is the decomposition of the tissues themselves. If this cause operates at all, it is only in the severest cases or in the last stages, where the strongest evidence in its favor is the irresistible wasting which characterizes these cases, in spite of the enormous food-consumption. In such event, the tissues would be split up into urea and sugar. And although emaciation may occur, because the albumen of the food ingested is not assimilated but passed directly through the economy, chiefly as urea, the comparative researches of Reich, Gaethgens, Pettenkofer, and Voit show that certain diabetics voided with the urine more nitrogen than corresponded with the nitrogen ingested.[*] If these researches are to be received, then must we admit that in certain very advanced cases the increased urea-excretion, as well as sugar-elimination, must be at the expense of the tissues.

Whether the urea is derived from the tissues or the food ingested, it is not impossible that the albumen is split up into urea and sugar as already suggested. For Dr. Haughton has shown that albumen, by the addition of small quantities of water, carbonic acid, and oxygen, furnishes the elements to produce glucose and urea in the proportion by weight of nearly

---

[*] Senator, loc. citat., p. 899.

five grains of glucose and one grain of urea. According to this theory, as Dr. Dickinson neatly puts it, "albumen is put into the mill, sugar and urea come out; and, like the flour and the bran, wax and wane together." According to this theory, such a splitting up takes place in health, but the sugar is not discharged in the urine as the urea is, because it is oxidized for the production of heat, or possibly converted into fat and stored up as such. And if the ratio of urea to sugar, as ascertained by actual analysis, be not very definite after allowing for the normal excretion of urea, it is questionable whether this fact should be admitted as an objection against the theory, for under any circumstances there must be a very great difference in the rate in which the two substances enter the blood, on account of the marked difference in their physical properties, as to osmosis, etc.

As regards *uric acid*, some observations tend to show that it is slightly diminished in diabetic urine. From my own observation I only know that, as a sediment, it is not infrequently present in diabetic urine.

Of the other constituents of the urine *sulphuric acid* is subject to its normal variations; *chlorine, phosphoric acid, lime,* and *magnesia* are said to be increased, phosphoric acid and lime especially so. Neubauer, Bœcker, Benecke, Vogel, Gæthgens, and Dickinson are all agreed to this. While admitting that most of the phosphoric acid, that combined with the alkalies, is derived from the food, Dr. Dickinson is inclined to think that the smaller part, that combined with the earths, is specifically increased by the disease. The phosphoric acid is derived from the disintegration of nervous tissue, while the lime which does not exist in nervous tissue is dissolved out of the other tissues by the acid in its transit.*

Of *abnormal* constituents, albumen is occasionally present. I have found it twice in seven cases. It is not generally large, but I believe that ordinarily it is the result of renal involvement, which is of the nature of congestive catarrhal nephritis,

---

* Dickinson, op. citat., p. 124.

though mild in degree. Thus might be accounted for the granular and hyaline casts I have found attending these small albuminurias, symptoms which, under other circumstances, would point to contracted kidney. Of course it is not impossible to have a contracted kidney accompany diabetes, but reasoning from the state of the circulation, epithelial changes are what we would expect, and autopsies, so far as they reveal organic changes, show them to be of the latter character. It is scarcely necessary to say that the urine may become albuminous from any of the causes of albuminuria independent of diabetes, as pus from pyelitis, cystitis, etc.

Inosit or muscle-sugar occasionally replaces the grape-sugar in diabetes, but more frequently accompanies it. Gallois* found it in five out of thirty-five diabetics.

Finally *alcohol* and *acetone* have both been found in diabetic urine, the former by Rupstein and Kuelz, and the latter first by Petters and afterwards by others. Acetone is believed to be derived from the decomposition of the diacetic acid of diacetate of ethyl, also present; alcohol partly from the same source and partly perhaps from the sugar. To acetone and alcohol is ascribed the vinous odor sometimes present in the urine, more particularly some time after it has been passed, though it may be immediately after it. A very simple *test* for acetone, discovered by C. Gerhardt, is a solution of chloride of iron, by which the urine is colored red.

Of the other secretions the *perspiration*, when present, frequently, although not invariably, contains sugar, sometimes a notable amount, as much as 6½ grains having been extracted by Fletcher from a piece of flannel three inches square, which had lain upon the skin 48 hours. The *saliva* has rarely been found to contain sugar independently of that which it acquires from the food. Whether the *gastric juice* ever contains it under similar conditions is disputed, but it has been found in effusions and exudations, as might be expected.

---

* Gallois, Comptes Rendus, 1863, I, p. 533; also De l'inosurie. Paris, 1864.

Duration.

Acute diabetes may be said to be unknown, and although cases are related by Becquerel, Wallach, Senator, Dr. Bence Jones, Roberts, and Dickinson, in which death followed within a very short time after its detection,—periods ranging from six days to six weeks,—yet in no instance can it be averred that the disease was of as short duration as it seemed, while in several there was reason to believe that it has lasted longer. It may be said then that diabetes mellitus is a disease almost invariably of long duration. A case of intermittent diabetes, under the successive observation of Dr. Prout and Dr. Bence Jones, lasted 16 years, and one was under similar observation of the last-named physician and Dr. Dickinson for 15 years. Lebert mentions a case which had lasted 18 years, the patient enjoying good health while on appropriate diet, but lapsing immediately to diabetism on an indiscriminate diet. The younger the age the shorter the duration and the more promptly fatal the result, while after middle age under treatment the duration may be indefinite.

The following table from Dr. Dickinson's work contains the duration of 100 cases collected by Griesinger, and alongside of it that of 25 which ended fatally at St. George's Hospital:

Duration.	St. George's Hospital.	Collected by Griesinger.
Less than ¼ year, ...	1	1
Between ¼ and ½ year, ...	2*	2
" ½ " 1 "	8	13
" 1 " 2 "	6	39
" 2 " 3 "	5	20
" 3 " 4 "	1	7
" 4 " 5 "	0	2
" 5 " 6 "	0	1
" 6 " 7 "	0	2
" 7 " 8 "	0	1
Undetermined, ...	2	12
	25	100

* In both these instances death occurred at six months.

From which it may be concluded that diabetes rarely proves fatal in less than six months after its accession, while few cases last more than four years. It must be remembered, however, that these statistics are drawn from hospital cases, which before admission were necessarily under much more unfavorable conditions than patients in private practice. Hence it is almost certain that the average duration of cases would considerably exceed that deduced from this table.

Complications.

The only complications of frequent occurrence in diabetes are the catarrhal pneumonias already referred to, which are rather to be considered a consequence. The same may be said of the boils and furuncles with which the skin is sometimes infested, and the occasional renal catarrh. Jaundice sometimes occurs, and having presented itself twice in the history of a case under my observation, can hardly be considered accidental, although I am at a loss how to account for it. Senator says that, when not an accidental complication due to a catarrh of the duodenum, it may result from compression of the biliary capillaries by the overloaded bloodvessels or enlarged gland-cells of the liver.*

Diagnosis.

The diagnosis of diabetes mellitus is very easy after it is once suspected. Unnatural thirst and copious diuresis should always suggest a chemical examination of the urine, but unfortunately these symptoms are not always present, and it is only by adopting a careful habit of testing the urine in all cases of disease, the least doubtful in their nature, that some obscure cases are detected. And although there are sources of error in testing for small quantities of sugar in urine, which are only overcome with considerable trouble, the quantities of sugar thus difficult of detection are not *usually* of clinical significance. Almost any one of the tests, therefore, which are found in the various manuals for the examination of urine,

---

* Senator, loc. citat., p. 912.

applied with ordinary care, will respond to such quantities of sugar. Such response being obtained, it may be concluded without hesitation that diabetes is present. In my own hands, no test for qualitative purposes has proved more delicate than Trommer's method of using the copper test.* Its ingredients are easily attainable, and there is no risk of error from changes to which Fehling's and Pavy's qualitative solutions are subject from age, while the latter are subject to the same sources of error, either from reduction by other substances than sugar, and to obscuration of the reduced oxide, which is occasionally held in partial solution by other constituents, as ammonia, creatin, etc.; thus making it difficult to assert positively that a reduction has taken place. In the latter instance a greenish or brownish-yellow results according as the quantity of copper solution added is large or small. All practical difficulties are removed by filtering the urine through animal charcoal, which removes the substances which interfere with the test either by reducing the copper or keeping it in solution; and often diluting the urine with water will be sufficient to secure an unmistakable reaction.

My method of using Trommer's test is as follows:

1. A drop or two of a (preferably weak—say 1 to 30) solution of cupric sulphate is added to the suspected urine, and then liquor potassæ or sodæ equal to half the total volume. On first adding the alkali there is immediately liberated, in addition to the earthy phosphates, a blue precipitate of hydrated cupric protoxide, *which, if sugar is present, is redissolved on adding more alkali*, producing a clear blue transparent liquid. If, on the other hand, no sugar is present, the fluid

---

* A reviewer of the second edition of my little book on the Examination of Urine, in the Medical Times and Gazette (London), was pleased to characterize my fondness for this test as old-fashioned. I do not think this view will be sustained by those who are in the habit of making very many qualitative testings of urine, using both Trommer's test and the quantitative solutions of Fehling, and I am glad to be able to quote the very explicit language of Senator, who says (article Diabetes, Ziemssen's Cyclopædia of Medicine, vol. xvi, p. 963): "Trommer's test deserves to be ranked foremost on account of its certainty and ready applicability."

will not be thus blue after the addition of the copper and alkali, but exhibit rather a turbid greenish hue. This, however, is not alone relied upon, but the *mixture is heated just to boiling*, and if sugar is present, a copious yellow precipitate of *hydrated* cupric suboxide takes place. This subsequently loses its water and becomes the *red suboxide* which falls to the bottom or sides of the test-tube, to which it often closely adheres. Occasionally the precipitate of earthy phosphates is so copious as decidedly to obscure the reaction. In this event they may be removed by filtration after adding the alkali and slightly warming the mixture, before adding the copper and further heating.

2. A second similarly prepared mixture of these ingredients should be made and set aside for from 1 to 24 hours without the addition of heat. If sugar is present a similar precipitate of suboxide of copper will take place. This repetition of the test is very important, since, according to Neubauer, the other organic substances which reduce the salts of copper do so only after long boiling. Hence also *prolonged boiling* should always be avoided.

When Fehling's or Pavy's solutions are used, the following method should well be found delicate:

A small quantity should be placed in a test-tube, diluted with about five times its bulk of water, and boiled alone for a few seconds. If the solution remains clear on thus boiling, add immediately the suspected urine drop by drop. If sugar is present in any quantity, the first few drops will usually cause the yellow precipitate, but if the reaction does not occur, the dropping may be continued until an equal volume of the urine has been added, when the mixture is again boiled.* If no precipitate occurs, sugar is absent. These tests are undoubtedly the most brilliant.

* One of the sources of error in this mode of testing is the prolonged boiling which a specimen is almost sure to receive. A few drops of urine are added to the test fluid, the mixture is then boiled, a few more drops are added, the mixture is again boiled, and thus a constant boiling is kept up until finally a reduction takes place, which may be from prolonged boiling alone.

If a precipitate occurs on boiling the *test fluid alone*, a new supply may be obtained, or a little more soda or potash may be added, the fluid filtered, and it is again ready for use. The precipitate referred to is a suboxide of copper, the result of a spontaneous reduction of the protoxide which sometimes occurs when Fehling's or Pavy's solutions are kept for some time. Boiling causes its precipitation, and hence the necessity of boiling a solution which has been kept for any length of time, before adding the suspected fluid. All possibility of such source of error may be avoided by keeping the solution of copper separate from that of the potash and potassic tartrate, and mixing them at the moment they are required for use.

In doubtful cases, also, urine passed two or three hours after a meal should be tested, as well as that passed fasting, for the former will often contain sugar when the latter does not.

In judging of the progress of a case of diabetes under treatment, it is not sufficient to test the urine qualitatively, but a quantitative determination of sugar must be made. This may be done by the volumetric processes described in the manuals for the examination of urine, but the simplest method is the fermentation method of Dr. Roberts. In this the specific gravity of the urine is taken before and after fermentation, and the difference in the specific gravity indicates the number of grains of sugar in each fluid ounce of urine. Then suppose the specific gravity before fermentation to be 1045, and after fermentation 1035, the quantity of sugar per fluid ounce is 10 grains.

Prognosis.

The prognosis of diabetes is generally considered unfavorable, but depends very much upon the age at which the disease makes its appearance, the time which has been allowed to elapse before treatment is instituted, and the treatment itself. Once thoroughly established early in life, or before 25 years of age, recovery would seem to be impossible, while even at this age, if treatment is instituted sufficiently early, much may be done to avert the end. Diabetes is a disease in which the expectant plan of treatment is disastrous. It is a disease which never gets well of itself, and always gets worse if not properly treated.

When the disease appears after middle life, is early recognized and promptly treated, it is ordinarily easily controlled; and although it is almost never safe to declare a case of diabetes absolutely cured, it does occasionally happen that recovery is so complete that the patient may be left to his own mode of living. As a rule, however, even cases which have apparently recovered have to keep a watch upon their diet, and should at intervals have their urine examined with a view to sounding, as it were, their condition.

When diabetes depends upon recognized nervous lesions the prognosis is altogether that of the lesion itself.

The cause of death is very frequently some intercurrent or consequent disease, as consumption. The extreme debility which ensues sooner or later is of such a character that there is no power of resistance, and a disease which would ordinarily be trifling, becomes, on this account, often a fatal one.

Treatment of Diabetes Mellitus.

I have already said that diabetes is a disease in which the expectant plan of treatment is disastrous, that it never gets well of itself, and that when left alone it almost invariably gets worse. The importance of a prompt and correct treatment is therefore evident.

From the discussion of the pathogeny of the disease it is also plain that at least an abatement of the most important symptom, glycosuria, may be expected by regulating the diet. Experience justifies such expectation, and it so happens that the abatement of this symptom is almost invariably followed by an abatement of all. The treatment, therefore, naturally divides itself into *dietetic* and *hygienic* and *medicinal*.

I. *Dietetic and Hygienic Treatment.*

By far the most important line of treatment, without which indeed no sufficient results have ever been attained, is the dietetic. This consists essentially in the elimination from the food of the patient of such articles as are readily convertible into sugar. It is generally acknowledged that in the early stage of the disease all the sugar which appears in the urine is derived in some way, whether directly or indirectly, from the saccharine and amylaceous alimentary principles; that the liver has lost its power to assimilate these, and they pass directly through the latter organ unconverted. Hence, if these be excluded from the diet and their place supplied by other assimilable articles, the symptom disappears, and the disappearance of this symptom seems to be, for the time being at least, the cure of the disease. For with it disappear also the frequent micturition, thirst, dryness, etc.

If it were necessary to select a diet absolutely free from sugar and starch it would indeed be restricted, as there are comparatively few articles of food thus constituted. Such are, however, meats of every kind, fresh or salted, including tripe, tongue, ham, bacon, and sausage; soups made from meat without flour; game, poultry, fish, oysters, lobsters, crabs, eggs, in every shape; butter and new cheese, oils and fats. Happily, however, it is not necessary to use articles absolutely free from the two baneful principles, and in this manner quite a variety of palatable articles may be added to the dietary. Among these are cream, curds, milk and buttermilk, all green vegetables, including spinach, endive, the green leaves of lettuce, dandelion, cabbage, coleslaw, brussels sprouts, cauliflower, broccoli, string beans, watercress, celery tops, asparagus tops, turnip tops, young onions, cucumbers, pickles, and olives. To these may be added unsweetened jellies (preparations of gelatin), and especially a variety of nuts, including almonds, walnuts, butternuts, filberts, pecan nuts, Brazil nuts, but not chestnuts; also, all acid fruits, as apples, oranges, lemons, strawberries, etc. Tea and coffee, with cream and without sugar, cocoa-nibs, but not chocolate, are permitted; also, all wines which contain little or no sugar, including claret, Burgundy, Rhine, and still Moselle wines, together with very dry sherry, unsweetened brandy, and whiskey; and gin when required. The carbonated waters, natural or artificial (the so-called soda-water of the shops), are pre-eminently suitable.

Water is to be allowed *ad libitum*, for water is the medium by which the sugar is carried out of the blood and tissues. Its supply should therefore be liberal, and with the diminished sugar-formation comes diminished thirst.

Beer, ale, porter, cider, and the fermented liquors generally, are not allowable because of the sugar they contain. They are less objectionable when fermentation is carried to a high degree, resulting in a more complete destruction of the sugar. This is the case with certain bottled lager beers and English ales.

It is not simply on account of the small quantity of sugar and starch contained in them which renders the vegetable substances named admissible, for many of them contain a good deal of sugar; but these sugars, unlike grape-sugar, are assimilable. Such are pre-eminently mannite, the sugar of manna, lactin or sugar of milk, lævulose or fruit-sugar, and probably, also, inosit or the sugar of muscle and the sugar of honey. Such is also *inulin*, a hydrocarbon and starchy principle found in the *inula helenium* or elecampane, but especially in Iceland moss. Hence, too, the impunity with which milk can often be taken by diabetics, although it contains from three to six per cent. of lactin. On this account, too, mannite may be used for sweetening tea and coffee where this addition is indispensable to the patient. Glycerin is also sometimes used for the same purpose, *i. e.*, as a substitute for sugar, but although less objectionable than sugar, it is not only theoretically unsuitable, but experience has shown it to be so; for glycerin is probably converted into glycogen or sugar in the liver, nearly two molecules of glycerin ($C_3H_8O_3$) being required to furnish one of sugar ($C_6H_{12}O_6$), or glycogen ($C_6H_{10}O_5$). Further, under the careful observation of Dr. Pavy,[*] it was noted that under the use of glycerin the urine increased from 3 and $3\frac{3}{4}$ pints to between 5 and 6 pints, and the sugar from 1100 grains to 3000 grains *per diem* in the course of three days. Its withdrawal

---

[*] Pavy, On Diabetes, London, 1869, p. 259.

was followed by a prompt fall in both the urine and sugar, a return to it by a second increase, and subsequent withdrawal by another decline. With the increase in the quantity of urine and sugar, came an increase in the thirst and discomfort; so that it would seem conclusive that the tendency, at least, of glycerin is to aggravate the symptoms, and its use, therefore, contraindicated. Mannite is therefore much to be preferred to glycerin, either for sweetening or to substitute the part of sugar in force-production.

It will be noted that not only all saccharine substances of animal or vegetable origin, and all vegetables largely composed of starch, as potatoes, rice, and corn, are omitted from the category of admissible articles, but that *bread*, and all preparations made of wheat, rye, rice, or corn-flour, are conspicuous by their absence. This is found to be a very important omission from the dietary of most persons, and very numerous have been the attempts to devise substitutes for it, with varying success.

Perhaps the best substitute for ordinary wheat-flour is the *gluten-flour*. It was suggested in 1841 by Bouchardat, and is made by washing the ordinary wheat-flour to free it from starch. Although this is not completely accomplished, a quite pure article of gluten may be obtained, containing, it is said, but one per cent. of starch, and some starch is necessary, if it is desired to make it into dough. It is perhaps best used in the shape of a mush or porridge, but it can be made into almost any of the various forms of bread and biscuit into which flour is made, and the ingenuity of bakers has been exercised in many ways to devise a palatable article. It is more nutritious than the bran-flour. As heretofore made, the salts have also been washed out by the process employed, but it is now made in such way as to retain the salts.* It is less irritating

---

* The Health Food Company, of 74 Fourth Avenue, New York, make a gluten of this kind by first removing the five bran coats, pulverizing the cleaned berry by the "cold-blast process," stirring the powder into ice water, precipitating the gluten, cellulose, and mineral matters, siphoning off the water, holding in suspension the starch, and drying out the precipitate.

to the bowels than the bran, and where this property, so often desirable, is objectionable it is more suitable than the latter.

Senator gives another method of getting rid of the starch and sugar in bread as tried by Vogel at the suggestion of Liebig. It consists in converting the starch into sugar by the action of diastase, and dissolving out the sugar thus produced. It is done by treating thin slices of bread with an infusion of malt. The bread is then washed, dried, and slightly toasted.

Another of these substitutes is the *bran-flour* deprived of its starch by washing. The bran itself is not wholly innutritious, containing, according to Parkes,* sometimes as much as 15 per cent. of nitrogenous matter, 3.5 per cent. of fats, and 5.7 per cent. of salts, although, in consequence of its indigestible character, it is probably not much availed of in nutrition. Moreover the salts are washed out in removing the starch which is mixed with it. But it is especially valuable in contributing a desirable bulk to the food of which it forms a part, and by its slightly irritant properties aids in maintaining a proper action of the bowels. These properties may be diminished, and the bran made much more suitable for its purposes by being very finely comminuted. It may be made by the ingenious cook into a variety of more or less palatable cakes. But sometimes, even when most carefully prepared, it is too irritant to be borne. Dr. Prout very early recommended, as a substitute for bread, a compound of bran, milk and eggs, which he declared not unpalatable.†

A purified gluten made by them is deprived of the cellulose walls of the cells in which the gluten-granules are held.

They also furnish directions for making gluten bread and gluten cakes of various kinds, as well as the gluten porridge, which is made by stirring the gluten into boiling water until thick enough, and then keeping up the boiling process for fifteen minutes. A little salt and butter are added at the close, to improve the flavor, and it may be eaten with milk or cream.

* Parkes, Practical Hygiene, 5th ed., Philadelphia, 1878, p. 222.

† A very carefully prepared *bran-flour*, as well as a *wheat-gluten flour*, is made by John W. Shedden, pharmacist, corner Broadway and Thirty-fourth Street, New York city.

The following are Dr. Camplin's directions for making biscuit of the bran-flour: To one-quarter of a pound of flour add three or four fresh eggs, one

In cases which do not require an extreme regulation of diet, I am in the habit of recommending the ordinary "bran bread" of the bakers, which is really bread made of unbolted flour, and contains the starch along with the gluten and bran; but the former is in much smaller proportion than in the bolted flour, and I note from its use an immediate diminution in the quantity of sugar excreted as compared with that under the use of the ordinary bread. From what I can learn of it, I have no doubt the "cold-blast whole wheat-flour would be better."*

Still another substitute for wheaten bread is the *almond food* suggested by Dr. Pavy. The almond is composed of 54 per cent. of oil, 24 per cent of nitrogenized matter known as emulsin, 6 per cent. of sugar, 3 per cent. of gum and *no starch*. Chemically speaking, it is therefore admirably adapted for diabetic food, and when the sugar and gum are extracted leaves nothing to be desired. The sugar and gum are removed by treating the powdered almonds with boiling water slightly acidulated with tartaric acid, or soaking the almonds in a boiling acidulated liquid, which may form part of the process for blanching. The boiling and acid fluid are necessary in order to precipitate the *emulsin*, which would otherwise emulsify the oil of the almond. I have had no experience with the almond food. Dr. Pavy speaks very highly of biscuits made of almond-flour and eggs, which, he says, go very well with a little sherry or other wine. He admits that they are found too rich by

and a half ounces of butter, and half a pint of milk; mix the eggs with a little of the milk, and warm the butter with the other portion, then stir the whole together well; add a little nutmeg or ginger or other agreeable flavoring, and bake in small forms or pattipans. The cake when baked should be about the thickness of an ordinary captain's biscuit. The pans must be well buttered. Bake in rather a quick oven for half an hour.

These cakes or biscuits may be eaten by the diabetic with meat or cheese for breakfast, dinner, or supper; at tea they require rather a free allowance of butter, or they may be eaten with curd or any of the soft cheese.

* This is another product of the Health Food Company, which contains the nutritious but not the innutritious portions of the bran. It is made by pulverizing the carefully cleaned wheat by a compressed cold-air blast, which strikes the wheat and dashes it to atoms.

some for ordinary consumption. A physician, himself a diabetic, with whom I am in correspondence, says he ate them but once and found them most unpalatable. He may, however, have hit upon a spoiled or imperfect article.*

Biscuits made of *inulin*, the starchy principle already referred to on page 280, were suggested by Kuelz.† Lichenin, or moss-starch, abundant in Iceland moss, is a variety of inulin and would be the material used for the purpose. Being very cheap it would be suitable on this account. Though a starch, it is, according to Kuelz, one of the assimilable starches already alluded to, of which small quantities, at least, do not increase the excretion of sugar. The biscuits are made with the addition of milk, eggs, and salt.

Under the head of dietetic treatment belongs the *skim-milk* treatment, of which Dr. A. Scott Donkin is the chief exponent and advocate. This treatment is based upon the view that lactose or sugar of milk is a material assimilable in diabetes, "and does not in the slightest degree contribute to the formation of sugar;" that in this respect it is even superior to casein, which, however, resists the sugar-forming process of the malady "to a degree immeasurably greater than any other albuminous alimentary substance, so that in all but the most severe and advanced or complicated cases it is complete."‡ With regard to casein, an albuminous substance, I presume no one will dispute, in general, the view taken by Dr. Donkin, although all may not agree with him in assigning to it the highest posi-

---

* Seegen recommends an almond food made as follows: Beat a quarter of a pound of blanched sweet almonds in a stone mortar for about three-quarters of an hour, as fine as possible; put the flour thus produced into a linen bag, which is then immersed for an hour and a quarter in boiling water, acidulated with a few drops of vinegar. The mass is then thoroughly mixed with three ounces of butter and two eggs; the yolks of three eggs and a little salt are added, and the whole is to be stirred briskly for a long time. A fine froth is to be made by beating the white of the three eggs, and added. The whole paste is now put into a form smeared with melted butter and baked by a gentle fire.

† Kuelz, Bieträge zur Path. und Therapie des Diabetes Mellitus, Marburg, 1874, Bd. i, s. 145.

‡ Donkin, Diabetes and Food, New York, 1875, p. 132.

tion in this respect. But as to lactose, the most prominent English physicians, among whom may be named Drs. Pavy, Roberts, and Dickinson, seem to be with singular unanimity opposed to him, while in their criticisms a tone of bitterness is not altogether wanting. In addition, however, to the results of his own observations,* Dr. Donkin is sustained in his views as to the assimilability of lactin by the modern German physicians and physiological chemists. It will be remembered that mannite, inulin, and lævulose are placed in the same category of assimilable carbohydrates by Kuelz. There is, further, considerable clinical testimony in favor of this treatment from others than Dr. Donkin who have tried it, and although I have as yet had insufficient personal experience with a pure skim-milk treatment as directed by Dr. Donkin, I know of at least one case in which the patient, a very intelligent physician, declares himself cured by a rigid adherence to it, and should opportunity present, where the ordinary dietetic measures failed to be followed by sufficient results, I would try it with no little hope of success. The instances in which I have tried it were hospital cases, where I met the usual difficulty, an unwillingness on the part of patients to submit to it; and in private practice I have up to the present time been satisfied with the results of the ordinary dietetic plan.

In what may be regarded with our present knowledge of the subject, and without pretension to too precise accuracy, as the first stage of diabetes, the dietetic measures above indicated are usually followed by the most prompt and decided results, occasionally by the permanent removal of all symptoms, at others by a continued absence of them so long as a watchfulness over diet is maintained. In a more advanced stage of the disease, in which a more rapid emaciation and loss of strength show themselves, such a regimen is followed by a decided dimi-

---

* For an account of Dr. Donkin's experiments the reader is referred to Dr. Donkin's communications to the London Lancet from 1868 to 1875; to his work on The Skim-milk Treatment of Diabetes, London, 1871, and the Transactions of the Clinical Society of London, vol. vii, 1874.

nution in the quantity of sugar excreted, but it fails to disappear altogether, and a more rigid elimination of saccharine and amylaceous articles must be attempted. Although it is commonly believed that sooner or later even the albuminous principles of food fail to be assimilated but largely reappear as sugar in the urine, Senator calls attention to the fact "that the formation of sugar or glycogen out of pure albumen (fibrin, white of eggs) has not thus far been proved, but is yet quite doubtful, and that still less has any diabetic been observed to pass sugar while using a diet *absolutely* free from sugar and the elements which form sugar or glycogen; that is to say, a diet consisting only of albumen, the necessary salts, extractive matters and drink."*

The importance of this statement will be further appreciated when it is remembered that even a so-called exclusively meat diet does not consist of albumen alone, but contains also glycogen, as well as undoubted glycogen producers and hydrocarbons in the shape of gelatin, glycerin (in the fat), and inosit. It will be remembered that the experiments of Salomon go to show that glycogen appears in the livers of animals fed on fats alone, the glycerin therein contained being probably its immediate source, and that Dr. Donkin's observations go to show that in aggravated cases of diabetes the ingestion of fat is followed by increased elimination of sugar; so that it is probably impossible to secure a thoroughly non-glycogenic diet, and it is questionable whether, even if it were possible, other urgent needs of the economy would be sufficiently supplied by it. On the other hand, it is evident that further observation and experiment are still necessary on the matter of diet in diabetes, and that valuable additions to our knowledge may be expected therefrom.

---

* Senator, loc. citat., p. 859. On page 957, however, Senator, admits that one albuminoid substance, *gelatin*, has been proved a producer of glycogen, and that even the chemical production of saccharine matters outside of the body has been achieved with it, though with it alone, and not with true albuminates.

In conclusion, the following summary of articles admissible for diabetics will be found convenient for reference:

Food and Drink.

Shell-fish.—Oysters and clams, raw or cooked, in any way, *without* the addition of flour.

Fish of all kinds, fresh or salted, including lobsters, crabs, sardines, and other fish in oil.

Meats of every variety except livers, including beef, mutton, chipped dried beef, tripe, ham, tongue, bacon, and sausages. Also poultry and game of all kinds, with which, however, sweetened jellies and sauces should not be used.

Soups.—All made without flour, rice, vermicelli, or other starchy substances or without the vegetables named below as inadmissible. Animal soups not thickened, beef tea and broths.

Vegetables.—Cabbage, cauliflower, brussels sprouts, broccoli, green string-beans, the green ends of asparagus, spinach, dandelion, mushrooms, lettuce, endive, coleslaw, olives, cucumber fresh or pickled, radishes, young onions, watercresses, mustard and cress, turnip tops, celery tops, or any other green vegetables.

Vegetables to be especially Avoided.—Potatoes, white and sweet, rice, beets, carrots, turnips, parsnips, peas, and beans; all vegetables containing starch or sugar in any quantity.

Bread and cakes made of gluten, bran, or almond flour, or inulin, with or without eggs and butter. Griddle-cakes, pancakes, biscuit, porridges, etc., made of these flours. No pastry permitted unless made of the admitted flours and without sugar.

Eggs in any quantity and prepared in all possible ways, without sugar or ordinary flours.

Nuts.—All except chestnuts, including almonds, walnuts, Brazil nuts, hazelnuts, filberts, pecan nuts, butternuts, cocoanuts.

Condiments.—Salt, vinegar, and pepper in moderate quantity.

Jellies.—None except those unsweetened. They may be made of calf's foot or gelatin, and flavored with wine.

Drinks.—Coffee, tea, and cocoa-nibs with milk or cream, but without sugar. Also milk, cream, soda (carbonated) water, and all mineral waters freely ; acid wines, including claret, Rhine, and still Moselle wines, very dry sherry. Unsweetened brandy, whiskey, and gin. No malt liquors except those ales and beers which have been long bottled and in which the sugar has all been converted into carbonic acid and alcohol.

II. *Hygienic Treatment.*

Next in importance to the dietetic is the hygienic treatment of diabetes. This consists in providing perfect ventilation, bathing, and attention to the skin, together with muscular exercise.

The diabetic should breathe the freshest and purest air. While the cases are not numerous in which embarrassed respiration results in glycosuria, there are undoubted instances in which this has occurred, as in croup and whooping-cough ; and it is well known that asphyxiated lower animals are apt to have glycosuria. And although the glycosuria thus resulting is probably reflex, it can hardly be expected that the diabetic should improve under unfavorable respiratory conditions. He should not, therefore, live, work, or sleep in a confined atmosphere, but secure the most perfect ventilation, spending much of his time out of doors, and sleeping in large, well-ventilated chambers, with windows open, etc. Especially should he avoid exposure to irrespirable gases. Attention to the skin, or skin-culture (Haut cultur), as it is termed by the Germans, is most important to the diabetic. He should bathe at least twice a week in tepid or hot water, on going to bed in winter, and on rising, or both on rising and retiring, in summer. He should groom his skin thoroughly daily, either after the bath or independent of it on the days on which he does not bathe. A tablespoonful or two of sodium carbonate to an ordinary bath is a suitable addition to the latter, softening the skin and facilitating its action by removing the effete epithelium.

Attention to other secretions, particularly the bowels, is of the greatest importance to the diabetic. It is probably partly on account of their action in this respect that the alkaline and alkaline saline aperient waters, as those of Vichy and Carlsbad, are of so much advantage. To those who visit these springs, a part of the advantage resulting therefrom is ascribable to the other favorable hygienic influences, such as rest, fresh air, and exercise, by which they are surrounded. Independently of these influences, however, I think there is reason to believe that the waters of Carlsbad and Vichy are of service to diabetics. And where their cost is not a consideration, from half a pint to a pint of Vichy and half as much Carlsbad, early in the morning, may be taken by the patient as an adjuvant in the treatment. The Vichy is a more alkaline water, containing 35 grains of carbonates to a pint, while Carlsbad contains but 11, but contains twice the proportion of chlorides, 8 grains to a pint, and nearly ten times as much sodium sulphate, or 19 grains to the pint; hence its more purgative quality. Since Carlsbad has the highest reputation it is more likely that it is through the action of the sulphates and chlorides on the liver rather than that of the alkalies they contain, that these waters are efficient. This is the more likely, as other alkaline waters nearly as rich as those of Vichy, and richer than Carlsbad in sodium carbonate, but without sulphate of sodium, are without reputation. The alkalies may, however, increase the effect, and are especially of service where there is acidity.

The only American waters known to me which approach these very closely as to both chlorides and sulphates, are those of Crab Orchard Springs in Kentucky, of which Sowder's Spring contains 25 grains of sulphate of sodium and magnesium, and 7 grains of sodium chloride to the pint, yet I am not aware that these waters have any reputation in diabetes. Other springs which approach them in the proportion of sulphates of sodium and magnesium are the Estill Springs and Harrodsburg Springs in Kentucky, and the Bedford Springs in Pennsylvania. The latter waters contain a little iron, which may be of advantage. The celebrated Saratoga Springs, in this country, have an undoubted action on the liver, probably through the chlorides they contain, which are in very large proportion, reaching in the Geyser Spring 70 grains to the pint, and in the Empire and Hathorn 63 grains to the pint. They contain no sulphates, according to Professor Chandler's analyses, but the carbonates are in considerable proportion, though much less than in the Vichy waters. In the absence of the Carlsbad and Vichy waters, I would use the purgative Saratoga waters, especially the Hathorn.*

Muscular exercise should be taken daily by the diabetic, both by walking and gymnastics. It is likely that glycogen is consumed to a degree in the muscles during their action, and that in diabetes there is an undue accumulation of sugar in the muscles is quite certain. The sense of muscular weariness so characteristic of diabetes is ascribed by some to this accumulation of sugar in the muscles. Dr. William Richardson,† of London, illustrates by his own case and that of others under his care, the undoubted benefit of exercise. It should be sustained regularly day by day, even in wet weather, care being taken to keep the feet dry, while it should never be carried to real fatigue. Dr. R. relates of himself, that at one time he had so little muscular power that he could not walk 100 yards without great fatigue, falling two or three times, and requiring always the greatest vigilance to prevent falling. He began to take exercise regularly two or three times a day, wet or fine.

---

* Some of the American waters which are most vaunted and advertised as useful in diabetes contain a surprisingly small quantity of either sulphates or chlorides, indeed of any of the ingredients which go to make a mineral water. Thus Bethesda water contains 1.7 grain carbonate of sodium and magnesium to the pint, .14 grain of chloride of sodium, and 1.1 grain of sulphates of sodium and potassium together, to the pint. Clysmic water about the same. The same may be said of the Poland and Capon waters. It is possible, however, that by drinking very large amounts of these waters, using them as table-waters and as substitutes for all other drinks, effects may be produced.

† Richardson, op. citat., p. 91.

He gradually gained strength, so as to walk five or six miles a day without fatigue. I cannot too strongly urge upon diabetics the following of Dr. Richardson's example according to his method.

Bouchardat (1835) and Trousseau early advised muscular movements, and recently (1875) Kuelz has strongly advocated them.

On the other hand, it is even more important that extreme fatigue should be guarded against, especially by those in advanced stages. Dr. Richardson reports two cases of sudden death after long journeys.

For a similar reason venereal excesses should be avoided, being peculiarly exhausting to one already weakened by diabetes.

III. *The Medicinal Treatment.*

Like all diseases in which treatment by drugs is relatively inefficient, diabetes has its full share of reputed remedies, most of which are useless. This dare not, however, be said of all.

In the first place I would call attention to the natural waters already referred to as adjuvants under the head of hygienic treatment. What was said of them there, might with equal propriety be said here.

At present, the first drug of which I make use in diabetes mellitus is *ergot*, not because it is invariably useful, but because I think I have been able to trace to it results more directly than to any other remedy. Moreover, the principle of treatment is a rational one. Ergot is supposed to produce contraction of the bloodvessel walls, and there is every reason to suppose it does. In diabetes there is dilatation of those of the liver. I give either the fluid extract in half-drachm doses, or preferably the gelatin-coated pills, of which one is equal to 30 minims of the fluid extract. Of these I give three a day, and continue the treatment for a month, if the stomach bears it. I have seen the sugar diminish and disappear under its use when the hospital diet was only modified by the addition of three pints of milk. At first the sugar reappeared after omitting the remedy, but finally disappeared not to return.

On the other hand, I have known it to be faithfully used for a month without evident results. I should be afraid, under ordinary circumstances, to rely upon it to the exclusion of the dietetic treatment.

It seems that *opium* was used for diabetes as early as the second century, by Archigenes. It was also used by Ætius of the ancients, and in the latter part of the eighteenth century and beginning of the nineteenth, by Rollo, Frank, Tommasson, and especially the English physician, Pelham Warren, in 1812.

It is certainly a useful agent in diabetes, and it is a remedy I would early use; but its use is united with disadvantages in the locking-up of the secretions which attends its use. Aperient remedies should therefore be used with it, and very suitable are the natural aperient waters, including the bitter waters, Friedrichshalle, Hunyadi Janos, Racoczy, Püllna, etc. Dr. Pavy* is a strong advocate of the use of opium, and has been astonished at the highly successful results. He has seen a patient entirely relieved under its use, and the use of it alone. It was given in increasing doses until the quantity reached *nine* grains in a day.† At first the glycosuria returned after its discontinuance, but on returning to its use the disease finally disappeared not to return. There was no restriction of diet whatever.

Morphia may, of course, be substituted for opium, and of the other alkaloids, *codeia* has been found useful by Pavy, Foster, and Image (quoted by Brunton). The latter gave it in doses of $\frac{1}{2}$ a grain three times a day, increased $\frac{1}{2}$ a grain every four days, until the patient took 5 grains three times a day. Reasoning from their physiological action, it would appear that opium and its alkaloids should be useful in cases where diabetes is the result of irritation.

---

* Pavy, op. citat., p. 275.

† McGregor (London Medical Gazette, 1837) allowed the quantity to reach, in one case, 60, and in another 90 grains within 24 hours. Whence it is inferred that diabetics bear opium without disadvantage in doses which would not be tolerated under ordinary circumstances.

Iodide of potassium has been found useful, and is a remedy to be early thought of.

Seegen has seen sugar entirely disappear under the use of 20 to 30 drops of *tincture of iodine* daily, but the sugar reappeared after discontinuing the remedy.

The *bromide of potassium* has been used by some with apparent advantage, and by others without effect.

Strychnia is one of the most useful adjuvants, acting as a tonic to the stomach and nervous system. It is best given in the form of the sulphate in acid solution, or it may be combined with *iron*, and especially with *iron* and *arsenic*. To the latter drug has been ascribed special influence in diminishing the formation of sugar, based upon the observation of Saikowsky, that glycogen diminishes in the liver of animals poisoned with arsenic. At any rate, the well-known pill of strychniæ sulph., gr. $\frac{1}{30}$, ac. arsenios., gr. $\frac{1}{30}$, and ferr. sulph., gr. 2, is here a very useful remedy. I would prefer, however, to give the strychnia in acid solution, and give the other remedies separately as required. Lenbe gave arsenic in doses of $\frac{1}{3}$ grain three times a day, with advantage.

To make up the deficiency in phosphates, which is the necessary result of the use of gluten bread, the preparation known as "Compound Syrup of the Phosphates" or Parrish's Chemical Food, is a good remedy. Each f$\mathfrak{z}$j contains 2$\frac{1}{2}$ grains calcium phosphate, 1 grain phosphate of iron, with fractions of a grain of phosphates of sodium and potassium, besides free phosphoric acid. It is administered in doses of a teaspoonful three times a day. The preparation of phosphoric acid and the phosphates* recently suggested by Professor William Pep-

---

* The formula is as follows:

R. Calcis phosphat., ...	grs.	iij
Magnes. phosphat., ...	gr.	ss.
Potas. phosphat., ...	gr.	iv
Ferri phosphat., ...	gr.	ss.
Ac. phosphoric (60 per cent.), . . .	℥vi	
Aquæ, q. s. ad	f$\mathfrak{z}$j.—M.	

Dose, a teaspoonful three times daily, diluted.

per, probably serves these purposes better, while it also fulfils the constant indication for a tonic.

Lactic acid was recommended by Cantani as a substitute for sugar, which in this disease is unavailable for the purposes it ordinarily serves in the economy. Adopting the view that the sugar ingested is converted into lactic acid in the liver, he would furnish the lactic acid already prepared to the latter organ, and thus, by giving it a rest, effect a cure. Senator, also, is inclined to favor its use, but on another ground. He believes that sugar is normally converted into lactic acid in the intestine, and that in diabetes the normal conversion may be interfered with.

The administration of lactic acid itself furnishes this important ingredient to the blood, where, as shown by Scheremetjewsky,* it is completely oxidized and becomes a force-producer. Therefore we must not look to a reduction in the quantity of sugar eliminated under its use, but regard it as a source of power, the patient getting the same advantage from it as the healthy individual does from sugar. They are said to gain in weight, become stronger, etc., and if they belong to the class in which the sugar disappears from the urine under an exclusively animal food, the disease may, by the employment of lactic acid at the same time, be wholly suppressed, and a condition of perfect health take place without excessive ingestion of food.† Cantani recommends that from 75 to 150 grains of the acid should be taken daily in from eight to ten fluid ounces of water. Larger quantities sometimes cause diarrhœa and pains in the joints, which disappear after its omission. A medical friend, who has apparently recovered from diabetes, used in conjunction with Carlsbad water and a pill of iron, quinine, and arsenic, 30 drops of lactic acid three times daily. My impression of lactic acid thus used as an adjuvant, is a good one, and I should be inclined to use it for the purposes named.

Glycerin has also been recommended by Schultzen as a substitute for sugar, on the supposition that sugar in health is de-

* Scheremetjewsky, Sächs. acad. Sitzungsb., 1869, p. 154.
† Senator, loc. citat., p. 999.

composed into glycerin and aldehyde of glycerin. But it has already been explained that glycerin is easily convertible into sugar, and that this conversion probably takes place in the intestines, while the experiments of Pavy and others have shown that glycosuria increases under its use. It is, nevertheless, not unlikely that in the early stages of diabetes, glycerin is assimilable, and it may be used in moderate amount for sweetening purposes, if the latter must be served, but I should hardly think, from what is now known of its properties and chemical composition, of using it as a remedy for the disease.

In view of the presence of glycerin in the fats themselves, in combination with the fatty acids, Senator suggests that the latter, viz., *oleic, palmitic, stearic,* and *butyric,* be administered in a separate form, on the same principle that lactic acid is substituted for sugar, that their force-producing power may be availed of. To this end he prescribed *soap* in the shape of pills, containing 2½ grains each, of which he directed four or five to be taken three times a day. In one instance he conjoined the treatment with cod-liver oil. In this there was no diminution in the amount of sugar excreted, but the patient gained in weight (11 ounces in fourteen days); and another thought there was a diminution in hunger and thirst. In both instances the diet was unrestricted, and the treatment was continued " several weeks " without digestive disturbances.

It might be expected that *cod-liver oil* would be a valuable remedy, or rather food, in diabetes. For while most experimenters deny that glycogen is ever produced under a diet of pure oil, those who claim that it is admit that, in the earlier stages at least of diabetes, all fats are assimilable. Cod-liver oil is one of these whose tonic and roborant properties have been too often tried to be any longer doubted. When well borne by the stomach, therefore, it may be early administered with the expectation that the general health and strength of the patient will improve while the excretion of sugar will diminish. The further indication for its use by the presence of phthisis in so many cases of diabetes, and the results of experience where phthisis is absent as well as when it is present, confirm the propriety of placing it in the category of eminently suitable adjuvants to the dietetic treatment.

The alkalies, and especially the alkaline carbonates, at one time enjoyed considerable reputation in the treatment of diabetes, after Mialhe claimed for them the power of destroying the sugar in the blood, and of neutralizing the volatile acids retained within the organism in consequence of the defective action of the skin. These views were subsequently refuted, but the carbonates continue to be used occasionally by many physicians, both in Germany and in England, but none seem to have had such favorable results as Dr. Pavy, who reports quite satisfactorily upon the use of potassium bicarbonate, in ten, fifteen, or twenty grain doses, in combination with aromatic spirit of ammonia. His experiments with large doses of sodium carbonate, four drachms a day; potassium acetate, six drachms; potassium citrate, six drachms; and rochelle salts, one ounce; were not, however, satisfactory. These were suggested by the results of his own experiments, in which he prevented the occurrence of glycosuria after operations on the sympathetic, by injecting sodium carbonate into the blood. It would seem, therefore, that when of service, it is by correcting the stomach and improving the digestion that the bicarbonates act.

In the essay already referred to by Dr. Dougherty, of New Jersey, he recommends a combination suggested by Dr. Whittingham, of potassium bicarbonate, sodium phosphate, salicylic acid, and sodium salicylate, made up with glycerin, compound tincture of cardamom and water, in the dose of $8\frac{1}{2}$ grains of the first, $2\frac{1}{2}$ of the second and third, and $4\frac{1}{2}$ of the fourth. My friend Dr. Andrew Nebinger informs me that he has used the salicylate of sodium in a single case with great advantage, in doses of 20 grains four times a day, after each meal and at bedtime, in equal parts of syrup and water.

Among the remedies which have acquired some reputation, is the *nitrate of uranium*. Attention was first called to it by Dr. J. Y. Dale,* of Lemont, Pennsylvania, who claims that

* Boston Medical and Surgical Journal, 1877.

it has been remarkably efficient in his hands in the treatment of diabetes. He prescribes it in doses of 1 grain three times a day, increased to three, if necessary, either in pill or in powder, or in solution, by the aid of a small quantity of nitric acid. He always uses it in connection with appropriate diet. Dr. H. A. Wilson has published in the *Medical Bulletin* (Philadelphia), for March, 1880, the results of treatment by this remedy, accompanied by a record of volumetric analyses; from which it appeared that there was a decided diminution in the quantity of sugar excreted. The dose was $\frac{1}{2}$ a grain increased to a grain three times a day, but the diet was also restricted. I have tried it in several cases, both with and without selected diet. In the former it was without effect, and in the latter there was none which I could not ascribe to the diet.

SECTION II.

DIABETES INSIPIDUS.

IF diabetes mellitus is considered to be an imperfectly understood disease, still more must this be said of the insipid diabetes or chronic polyuria, for even still less is known of its essential or remote causes.

The term is applied to any excessive secretion of non-saccharine and non-albuminous urine which has continued for a long time, and which is accompanied by extreme thirst.

• The condition, unlike diabetes mellitus, affects rather young persons, being rare in those over 50 years of age, relatively frequent in infancy, and most common between the ages of twenty and thirty, as is shown in the appended table from Senator's article in Ziemssen's *Cyclopædia*, of cases collected by Roberts, Strauss, and von der Heijden :

Age.	Roberts.	Strauss.	von der Heijden.
Less than 5 years,	7	9	2
5 to 10 years,	15	12	5
10 to 20 years,	13	} 57	19
20 to 30 years,	16		23
30 to 40 years,	} —		19
40 to 50 years,	} 15	} 7	9
50 to 60 years,	} —		6
60 to 70 years,	} 40		4
	70	85	87

Its contrast to diabetes mellitus in this respect, has been alluded to in connection with the former disease.   As to the *sex* of those affected, the same peculiarity is common to both ; it is much more frequent in males than females, two to three times as many of the former sex being affected by it as the latter.

Etiology and Pathology.

As to causes, the same uncertainty prevails as with diabetes mellitus. An examination of cases shows an association with a certain number of conditions, such as cerebral disease, including tumor of the brain, sunstroke, cerebro-spinal fever, blows on the head, and falls, exposure to cold, and drinking cold fluids, drunkenness, pregnancy, hysteria, emotion, hereditary influence, previous disease, etc., but this does not show causation. The proportion, however, of cases in which the condition is associated with brain diseases and injuries to the head, taken in connection with the fact of Bernard's discovery that puncture of the floor of the fourth ventricle above the diabetic centre, produces increased secretion of urine without glycosuria, but frequently with albuminuria, makes it very likely that central nervous irritation, however induced, is at the bottom of the symptom. Eckhard* confirmed this observation, and showed also that in dogs, at least, the channel through which the stimulus to increased secretion takes place is probably the spinal cord, at least as far as the seventh cervical vertebra, for section of the cord above this point abolished altogether the secretion of urine. On the other hand section of the greater splanchnic nerve caused an increased flow of urine on the same side, while irritation of the peripheral extremity of this nerve caused a diminution and even cessation of the secretion of urine.

Again, irritation of the floor of the fourth ventricle, after the splanchnic was cut, caused a further increase in the urine secreted. Whence it is evident that the splanchnics are inhibitory of the act of secretion of urine, while the excitor influence passes through another channel, for a certain distance at least, in the spinal cord. The experiments of Peyrani† upon dogs, cats, and rabbits go to show also that the excitor influence travels through the sympathetic, since electrical *irritation* of the cervical sympathetic lasting for several hours resulted in

---

* Eckhard, Beiträge zur Anatomie und Physiologie, iv, v, and vi.
† Peyrani, Comptes Rendus, 1870, i, p. 300.

an *increased* excretion of both urine and urea, while *section* resulted in a decidedly *diminished* secretion of both.   In a word,
the channel through which increased secretion of non-saccharine urine takes place is the same as that through which saccharine diabetes results, the difference being that in the latter
sugar is added, while in the former it is absent.

The immediate cause is a dilatation, first, of the arterioles,
and then of the capillary bloodvessels of the kidney, through
whose thinned and stretched walls the water osmoses readily,
so that the whole act is probably one of increased filtration,
although it has been suggested and deemed not improbable by
Eckhard that at least in polyuria from irritation of the floor
of the fourth ventricle, there is an excito-secretory influence exerted upon the gland-cells, increasing their activity, and in consequence the secretion of urine.   But no nerves have ever been
traced to the *cells* of the kidney, and while the cells are probably active in separating some of the organic constituents of
the urine, there is every reason to believe that the water, at
least, is simply filtered out.

Hence it may be inferred that diabetes insipidus is always
the result of some irritation, either direct or reflex, of this centre
in the medulla oblongata, or of the sympathetic ganglia in the
abdomen, a conclusion which is confirmed by the fact already
mentioned, that the largest number of cases associated with any
single condition is found in connection with diseases and injuries of the brain, and still more by the fact that in a considerable number of cases the lesion has been found to be in the
medulla oblongata, or the floor of the fourth ventricle.   It
would seem to differ also from diabetes mellitus in having
purely this mode of origin.   That is, it can never be said to
result from purely local hepatic or intestinal derangements, as
there seems every reason to believe may be the case with diabetes mellitus.

Morbid Anatomy.

The essential morbid anatomy of diabetes insipidus would be
the lesions of the nerve-centres or sympathetic ganglia which

may underlie the symptoms. But as these are often undiscoverable, or at least indefinite, it is impossible to describe them.

As a secondary phenomenon and consequence, rather than an essential morbid lesion, is an atrophy and degeneration of the renal substance, which may be trifling or have proceeded to a degree of complete sacculation of the organ. Out of eleven cases in which autopsies had been made, collected and reported by Dr. Roberts in his work on *Urinary Diseases*, four presented this alteration; in one, the glandular tissue was entirely wanting. In a fifth, multiple abscesses were found in one kidney, the other was hyperæmic. In a sixth, there was a tubercular tumor in the left hemisphere at the border of the longitudinal sinus, and another in the cerebellum. In the seventh and eighth cases, fatty degeneration of the nervous tissues of the walls of the fourth ventricle was present; in the eighth, there was also great vascularity of the kidneys. In the ninth, there was a glio-sarcoma as large as a walnut in the floor of the fourth ventricle, filling the entire cavity. In the tenth, there was miliary tubercle at the base of the brain, near but not in the fourth ventricle, and on the upper surface of the cerebellum. In an eleventh, there were no changes in the kidney, except congestion of the Malpighian bodies.

It will be noticed that nervous lesions when present are found more frequently in the vicinity of the base of the brain.

Symptoms of Diabetes Insipidus.

These are not numerous. The enormous secretion of urine, of almost spring-water-like clearness and of specific gravity often as low as 1003, is the most conspicuous symptom, but more annoying, probably, is the *extreme thirst* which always attends it.

These may be said to include all the essential symptoms, others which may or may not be present being rather their consequence. Very constantly among them are *dryness of the skin* and absence of perspiration. The health is otherwise often perfect.

Occasionally there are *derangements of digestion*, and sometimes also the appetite is large, as in diabetes mellitus. Always, the effect of increased discharge of water by the kidneys is to increase the rapidity of tissue-metamorphosis and to wash out thoroughly the urea and extractives from the blood and tissues, so that it might be expected that there would be some increase of appetite from this source.

These symptoms may occur suddenly in the midst of apparent health, or they may supervene upon or be superadded to others, chiefly of a nervous character, which may be the result of the same nervous lesion causing the polyuria. Such symptoms are headache, restlessness, irritability, more rarely convulsions, delirium, paralyses, indeed any one or more of the great variety of symptoms which result from organic or functional nervous disease. Sometimes these symptoms succeed upon the polyuria, or are increased by it. It is certain that the milder nervous symptoms are sometimes the result simply of the inconvenience and annoyance caused by the two cardinal symptoms, polyuria and thirst. The patient is kept busy, as it were, night and day, in passing water. It is not surprising that such a patient should be fretful and irritable, and that sooner or later his health should be broken if the symptoms are not relieved.

In addition to the symptoms detailed, there are said to occur at times *dryness of the tongue, epigastric and lumbar pains, diarrhœa, debility, impairment of mental faculties and sexual function*. The debility is sometimes extreme. In some instances there is the most extraordinary *tolerance of alcoholic drinks*, while at others there is an *exaggerated susceptibility* to their influence. A very *slight lowering of the body temperature* has been observed, amounting, however, to but a few tenths of a degree, and is never below 97° F. This slight lowering has been ascribed to the refrigerant effect of cold drinks, which are constantly ingested. The tendency to boils observed in mellituria is not found in diabetes insipidus. In advanced stages of the disease *œdema* of the ankles sometimes occurs.

The *duration* of the condition is very various. Sometimes it continues through life with no inconvenience except that from the constant diuresis and thirst, and no disadvantage when the latter is supplied. Dr. Willis records a case lasting fifty years. On the other hand, it is seldom of brief duration; indeed, there is needed a certain chronicity in order to admit it in the category of diseases. One case is reported as terminating fatally in seven weeks. Under prognosis will be found some further information as to duration, but it may be said, in general, that most cases which terminate unfavorably and most which recover completely, do so within a year.

It has *no complications* except such as are its cause or its results. Among the latter is occasionally dilatation of the pelvis of the kidney and atrophy of the latter, due to pressure of the accumulated urine, and resulting in a sacculated condition of the organ. Its symptoms are almost always influenced, and sometimes even cut short, by intercurrent disease, especially of a febrile character, or even by a profound physical impression, as long-continued suppuration after a blister. Thus Willis[*] cites the case of a young man who had thirst and diuresis up to 18 years of age. At that time he was attacked with pleurisy, and a blister was applied which suppurated for 25 days. With the healing of the blister disappeared both the pleurisy and the diabetes. Dr. Roberts[†] cites a case which had existed "in intensity" for 18 years, in which an attack of acute rheumatism treated with nitrate of potash suspended it completely. Kuelz, Charcot, and Lacombe record cases in which temporary improvement occurred during varioloid, and Senator[‡] of similar improvement during pneumonia, and again in the same man during erysipelas. On the other hand, Dickinson[§] records a case in which scarlatina supervened without effect on the symptoms. It has been known to occur during pregnancy and to disappear with it, while in other instances it has been uninfluenced by the latter condition.

* Willis, op. citat., p. 24.
† Roberts, Urinary and Renal Diseases, 3d Am. edit., 1879, p. 202.
‡ Senator, op. citat., 1031.
§ Dickinson, op. citat., p. 205.

Physical and Chemical Characters of the Urine.

As to the quantity of urine passed, it is enormous, exceeding often the amount passed in saccharine diabetes. As many as 43 liters (90 pints) are recorded by Trousseau, and one-fourth this quantity is common. It has been said, even, that the quantity secreted sometimes exceeds that ingested, but this is impossible, at least for any length of time, unless water is absorbed from the atmosphere, which is not likely. In point of fact the water excreted is always a little less than that ingested, either as drink or in the solid food.

As the quantity of urine excreted increases, or its normal acidity diminishes, its color disappears and its specific gravity declines. In one case under my care the specific gravity was scarcely 1001, while the urine in moderate bulk was absolutely colorless. Again, a faint greenish tinge is exhibited by the urine in bulk.

As to the other constituents of the urine, it may be said in general that they are all increased, except possibly uric acid. Thus the *urea* is increased to three and even four times its normal amount,—70 and 72 grams (1080.10 and 1110.96 grains) are recorded by Senator, in an adult, and 30 grams (462.9 grains) by Dickinson, in a child of 8 years. It has already been explained how this naturally results from the washing out of the tissues. In a case reported by Da Costa* the urea was diminished.

With regard to *uric* acid, its estimation is difficult, on account of the extreme dilution of the urine, but analyses would go to show that it is diminished, which is what one would expect where the urea is so much increased, oxidation being more complete.

Sulphuric and *phosphoric* acids are both increased, and especially, according to Dickinson, the combination of phosphoric acid with the earths, lime and magnesia.

---

* Transactions of the College of Physicians of Philadelphia, 3d series, vol. i, 1875, p. 139.

The same is true of the *chlorides*, which are increased for the same reason as urea, sulphates and phosphates.

Of abnormal constituents, *inosit* has been found, and it is said *albumen* very rarely, but care should be taken not to confound the polyuria with small albuminuria of a contracted kidney with an albuminous polyuria in which there is no organic disease of the kidney.

Unlike the urine of saccharine diabetes, that of diabetes insipidus rapidly decomposes, becomes putrescent, and at the same time turbid with bacteria. A further peculiarity of the secretion of urine is that an increase does not rapidly follow the ingestion of fluid as is the case in health, but ensues at a an interval, and is kept up during a longer period. This is ascribed to the *constant* dilatation of the renal capillaries which is supposed to exist in diabetes insipidus, as compared with the *alternate* dilatation and contraction which occur in health in the reflex accommodation which is always taking place between the ingestion of fluids and their secretion by the skin and kidneys. In health the ingestion of an undue amount of fluid is promptly followed by dilatation of the renal or cutaneous capillaries or both, and the transudation of an increased amount of urine or perspiration or of both. If, on the other hand, the quantity of liquid ingested is small, these two sets of capillaries remain contracted and the water is retained in the economy. In diabetes insipidus, on the other hand, the renal capillaries are always dilated, and are always therefore in a condition to permit the transudation of water, while they in turn cannot respond as promptly to the ingestion of fluid as they do in health. A longer time, therefore, intervenes before the increased secretion takes place, while it is also longer kept.

Some of the accounts published as to the quantity of water consumed and excreted are almost incredible, yet they seem well authenticated. The following illustrative instances are condensed from Dr. Willis's work on *Urinary Diseases*:* A small artisan, 55 years old, had had constant thirst with commensurate diuresis since he was 5 years of age. From

---

* American edition, Philadelphia, 1839, p. 23.

the age of 16 he had drank, on an average, no less than two pailfuls daily. While in the Hôtel Dieu, to which he was admitted for an injury of the knee, he drank on an average 33 pints of water every day, often swallowing two liters or about two quarts at a draught. He passed daily about 34 pounds of urine and 1 pound of fæces. He otherwise enjoyed good health, and was the father of several children.*

Again, a Frenchwoman, aged 40 years, had been afflicted from her birth with a drought beyond example. She drank every day nearly two pails of water, and was eventually driven from home by the ill-treatment she received in consequence of this expensive habit. At 22 she married a cobbler. She drank four pailfuls a day, and became the mother of 11 children, drinking more when she was pregnant and least when out of health. When 40 years old she was examined by a scientific commission, and drank in the presence of its members 14 quarts of water within ten hours, and voided 10 quarts of nearly colorless urine.†

Dr. Dickinson reports the case of a farmer, 51 years old, in good general health, and equal to severe farm labor, who usually drank a quart of water at a time, and repeated the draught sixteen or eighteen times in the day and night, passing about as much urine as he drank water. In one night, under observation, he passed between 5 and 6 quarts of urine without sediment.‡

These cases show also the extreme duration of some cases, and the otherwise excellent health enjoyed by them. Very little serious disturbance seems to result as long as water is supplied to quench the resulting thirst. In extreme cases patients have even been known to drink their own urine. Dr. Dickinson had two cases which did so.

An extraordinary flow of *saliva* was met in one instance by

---

* This case was reported by M. Boissat, in the Recueil de Sedillot, tome lxxx, p. 164.

† This case was reported by Mr. Maxwell to Dr. S. F. Simmons, who published it in Medical Facts and Observations, vol. ii, p. 73. It was las seen in 1791.

‡ Also from "Medical Facts and Observations," vol. ii.

Kuelz* along with polyuria in a hysterical girl of 18 years, from whom as much as 525 cubic centimeters (17.72 ounces) were collected in twenty-four hours, while the quantity ranged during four months from 360 to the former amount. The quantity of urine passed during this time ranged from 6000 to 7800 centimeters (200 to 260 ounces). The increased flow of saliva is explained by the fact that in some of the experiments of Eckhard,† Loeb‡ and Gruetzner,§ puncture of the medulla oblongata was followed by ptyalism.

Diagnosis.

The diagnosis of diabetes insipidus is very easy. The persistent thirst, polyuria, and absence of sugar from the urine are pathognomonic. It is simply necessary to be sure of these, to make the diagnosis positive. The only possible error is mistaking the polyuria of chronically contracted kidney of interstitial nephritis for that of diabetes insipidus. This I have known to occur from overlooking the presence of a very minute quantity of albumen. In addition, however, to the fact that a careful examination for albumen will discover it in the urine of contracted kidney, the quantity of urine is never inordinately large, nor is the thirst extreme; so that it would seem only necessary to mention the possibility of such an error in order to avoid it

Prognosis.

It is extremely rare for a case of diabetes insipidus to terminate unfavorably unless there have been also present symptoms pointing to serious nervous lesion, that is, it is not fatal by the virulence of any symptoms peculiar to it. On the other hand, recovery is almost as infrequent as death. According to Dr. Roberts, of 67 cases collected, 16 are reported as complete recoveries, and 14 ended fatally, nearly an equal proportion. The remaining 37 were still in progress. In cases

---

* Diabetes Mellitus and Insipidus, Marburg, 1875.
† Eckhard, Beiträge zur Anat. und Physiol., iv, p. 191.
‡ Loeb, Eckhard's Beiträge, v, p. 1 ; and Dissertation, Giessen, 1869.
§ Gruetzner, Pflüger's Archiv, vii, p. 552.

of recovery or death, the duration is comparatively short. Of the 16 recoveries, in 9, the duration was less than a year; in 1, four years; in 2, eighteen and nineteen years, and in the remainder some years. Of the 14 fatal cases, 9 terminated in less than a year, 1 *in seven weeks*, 2 in two months; the other 5 survived for periods varying from eighteen months to twenty years. Of the 47 cases in progress only 5 had continued for a year or under. The remainder had continued for periods ranging from something over a year to fifty-nine years.

These results seem to be tolerably independent of treatment. It may be said, therefore, that, as a rule, cases that last more than a year are apt to continue, but ordinarily only require to be furnished with an abundance of water to keep them tolerably comfortable. According to Dr. Dickinson, cases from drunkenness are very apt to run a severe and rapid course, usually terminating fatally within a few months, and one terminated thus in two months.

The disease appears to me altogether less serious than diabetes mellitus. This I infer to be the opinion of Roberts and Dickinson. It certainly is that of Senator, who says "it is rather a troublesome than a dangerous complaint." But Trousseau and Da Costa are inclined to consider it more serious than diabetes mellitus.

Treatment.

The treatment of diabetes insipidus would naturally resolve itself into the treatment for the disease of which it is the symptom, than of the symptom itself; but as the former is very frequently undiscoverable it must consist mainly of efforts to diminish the secretion of urine, and with it the thirst.

First, it is generally conceded that there should be no restriction in the drinking of water or other harmless fluids, for the diuresis is not so much caused by the large ingestion of water as the thirst is caused by the diuresis. To cut off the supply of drink must therefore result in a draining of the tissues to their disadvantage, and a corresponding increase in the thirst. It should be mentioned, however, that one or two instances are reported wherein improvement seems to have resulted from such restriction. To relieve the diuresis is therefore to relieve the thirst. Rational remedies to this end would seem to be _astringents_ and other drugs which tend to produce contraction of the capillaries and arterioles. But experience does not seem to sustain our expectations in these respects, probably for the reason that the lesion which causes the capillary dilatation resides elsewhere than in the kidneys themselves. Da Costa* reports a case of recovery from the use of _ergot_ in drachm doses of the fluid extract three times daily, increased after ten days to two fluid drachms as often. I myself† have found the symptoms to subside under the use of gallic acid after I had failed with full doses of ergot. Reasoning from the reputed action of ergot on the capillaries of the nerve-centres as well as those of the kidney, we would expect it to be the more efficient remedy, and from my present knowledge of the subject I should begin treatment with it rather than any other remedy of which I know, although Dickinson says that "remedies designed to restrain the urinary secretion seldom fail to do harm." So far as they may have any effect in disturbing digestion I would also discard them. Among the remedies to be cautiously used on this account is _opium_, which has had some reputation.

Trousseau and Rayer claimed extraordinary results from the use of _valerian_, the former using the extract in enormous doses,—two and a half drachms a day, which was increased to an ounce daily in one instance. Rayer used the powdered valerian and the valerianate of zinc, giving the latter in pills in gradually increasing doses until 20 grains a day were given.

Reasoning from the effect of intercurrent disease and powerful physical and nervous impressions, Roberts suggests a large _blister_ at the nape of the neck or epigastrium, according as the associated symptoms and the anamnesis point to the nervous

---

* Da Costa, Diabetes Insipidus and its Treatment by Ergot, Transac. College of Physicians, Philadelphia, 3d ser., vol. i, 1875, p. 139.

† Tyson, Case of Diabetes Insipidus Treated by Ergot and Gallic Acid, Transac. College of Phys., Philadelphia, 3d ser., vol. ii, 1876, p. 180.

or digestive system, a suggestion which may be acted upon with advantage.

Recently the use of the *constant galvanic current* has been recommended, and in cases of spinal lesion may be expected to be of advantage. Seidel and Kuelz have both used it with good results. The former applied one pole of a "strong battery" over the loins near the spine, and the other as deeply as possible over the hypochondrium, upon each side daily for five minutes. In eight days the urine fell from 5957 cc. to 4600 cc. per diem, in three weeks to 2300 cc., and the next month 1904 cc., while the weight of the body increased nine pounds.

Kuelz applied one pole of a battery of 30 to 40 cells as high as possible in the nape of the neck, and the other to the loins or epigastrium, the best results being apparently obtained with the positive pole to the, nape of the neck, and the negative first to the loins for four minutes, and then to the pit of the stomach for four minutes.

Tonics and *nervines*, such as strychnia, iron, arsenic, salts of quinia, cod-liver oil, etc., are appropriately added to the treatment with a view to sustaining the strength of the patient, which is apt to give way. To these are to be added fresh air, sea air, exercise, and all possible favorable hygienic influences.

Senator says: "Since diabetes insipidus is rather a troublesome than a dangerous complaint, it is advisable in the lighter cases to avoid the administration of drugs, and to recommend to the patients only a *careful attention to the skin*, warm clothing, warm baths, frictions, etc., in order to divert a portion of the stream of fluid from the kidneys to the skin." He also advises, in addition, in severer cases, to quench the thirst, not by excessive drinking, but by bits of ice and acidulous fluids. I have already expressed my preference for a treatment in which, under ordinary circumstances at least, the supply of water should be unlimited.

Among other remedies which have been recommended are, in addition to opium and its alkaloids, acetate of lead, tannin, digitalis, belladonna, bromide of potassium, iodide of potassium, iodide of mercury, camphor, jaborandi, lime-water, bitartrate of potassium, etc. I have had no experience with any of them.

INDEX.

Aladoff and Cyon, experiments producing glycosuria, 235
Albumen, cause of transudation of, 58; tests for, 50; to indicate approximately quantity of in urine, 54; quantitative estimation, 50
Albuminoid disease of kidney, 148
Albuminous nephritis, 85
Albuminuria, 50–61
  its mechanism and production, 50
Amyloid kidney, 148
Asthma, uræmic, 105
Author's classification of kidney diseases, 83
Bartels's classification of kidney diseases, 82
Bernard, theory of diabetes, 230–233
Blood-casts, 62
Blindness in acute uræmia, 104; in diabetes mellitus, 263
Bright's disease, acute, 83, 85
      chronic, 84
      classification of, 79
Brunton, on diabetes, 235–240
Casts. See Tube-casts
Catarrhal nephritis, acute, 85
      chronic, 124
Charcot, views on classification of kidney disease, 82
Chronically contracted kidney, 165
Cirrhosis, renal, 165
Cirrhotic kidney, 165
Classification of Bright's disease, 79
Cohnheim's experiments upon inflammation, 58
Coma, diabetic, 265
Contracted kidney, 165
Connective tissue of the kidney, 37
    Beale's views on, 38
    Beer's views on, 37
    Goodsir's views on, 37
    Henle's views on, 37
    Johnson's views on, 38
    Key, Axel, views on, 40
    Kölliker's views on, 39
    Ludwig's views on, 39
    Schweigger-Seidel's views on, 39
Croupous nephritis, 80, 81, 85
Cyanotic induration of the kidney, 80, 81, 219
    diagnosis of, 222; etiology and pathogeny of, 219; morbid anatomy of, 220; prognosis of, 222; symptoms of, 221; synonyms of, 219; treatment of, 223
Cyon and Aladoff, experiments producing glycosuria, 235
Depurative disease of kidney, 148
Desquamative nephritis, acute, 85
    chronic, 124
Dickinson, classification of Bright's disease, 83
    morbid anatomy of diabetes mellitus, 257
Diffuse nephritis, 124
Diabetes, 227
Diabetes insipidus, 298
    diagnosis of, 307
    etiology and pathology of, 299
    frequency of, 298
    morbid anatomy of, 300

Diabetes insipidus, prognosis of, 307
    symptoms of, 301
    treatment of, 308
    urine in, 304
Diabetes mellitus, 227
    blood, the alterations in, 266
    causes of, 229, 250
    complications in, 274
    course of, 259
    Cyon's and Aladoff's experiments producing, 235
    death-rate from diabetes, 228
    diagnosis of, 274
    Dickinson's description of the essential morbid anatomy of, 257
      views on diabetes, 242
    dietetic treatment of, 278
    Donkin, Dr. A. Scott, on treatment of, 284
    duration of, 273
    food and drink in, 287
    hygienic treatment of, 278
    pathology and pathogenesis of, 229
    prognosis of, 277
    Schiff's experiments producing glycosuria, 233
    symptoms of, 259
    treatment of, 278
    treatment, medicinal, of, 291
    urine, alterations in, 267
Diabetic coma, 265
Dickinson, classification of kidney disease, 83
Eckhardt on diabetes, 235, 237, note.
Fatty and contracting kidney, 124, 128
Frerichs on the unity of Bright's disease, 80
Glomerulo-nephritis, desquamative, 96
    glycosuria. See Diabetes.
    abnormal alimentary, 243
    normal alimentary, 242
Gouty kidney, 165
Granular degeneration of kidney, 165
Granular kidney, 165
Heidenhain on structure of the kidney, 26
Heller, bacteria in bloodvessels of kidney in pyæmia, 91
Heyl, A. G., retinal changes in diabetes, 261
Hyperæmia of the kidney, acute, 85
    passive, 80, 84, 219
Interstitial nephritis, 165
    causes of, 165
    complications of, 182
    connective tissue, overgrowth of, in, 173
    cysts, origin of, in, 173
    diagnosis of, 183
    duration of, 182
    etiology of, 165
    minute structure of kidney in, 172
    morbid anatomy of, 170
    prognosis of, 183
    symptoms and course of, 177
    symptoms of, 183
    treatment of, 183
    suppurative nephritis, 200
Johnson, views on kidney disease, 83
Kidney, cirrhotic, 165
    contracted, 165
    gouty, 165

Kidney, granular, 165
    large white, 124, 125
    large contracting, 124, 128
    fatty and contracting, 124, 128
    surgical, 200
Kidney, normal, 17
    anatomy of, 17
    arrangement of uriniferous tubules, 21, 23
    bloodvessels of, 32
    capsule of, 17, 40
    connective tissue of, 37
    glomerules of, 30
    epithelium of tubules, 25
    Heidenhain, on structure of, 26
    Klein's view on structure of tubules, 27, 28
    lymphatic vessels of, 40
    Malpighian capsules of, 20
    Malpighian tufts or glomerules, 24, 30, 33
    medullary rays, 18, 19, 20, 44
    methods of studying histology of, 42
    nerves of, 41
    pyramids of Ferrein in, 44
    preparation of sections of, 45
    spiral tubes of Schacowa in, 20
    uriniferous tubules of, 20
    vasa afferentia, 33
    vasa efferentia, 33
    vasa recta, 35
    weight of normal, 17
    passive congestion of, 219
    Klebs's classification of kidney disease, 81
    bacteria in suppurative nephritis, 201, 202
Lafont on diabetes, 238, note.
Langhans's views on changes in the glomerules of the kidney in disease, 94
Lardaceous disease of the kidney, 84, 148
    complications of, 159
    diagnosis of, 160
    Dickson's, Dr., views on, 157
    duration of, 159
    etiology of, 148
    morbid anatomy of, 149
    prognosis of, 162
    treatment of, 163
    suppuration, as a cause of, 148
    symptoms and clinical history of, 157
Lipæmia, intraocular, in diabetes, 261
Nephritis, acute, 85
    acute diffuse, 85
    chronic diffuse, 124
    interstitial, 165
    interstitial suppurative, 200
    parenchymatous, acute, 85
    parenchymatous, chronic, 124
    suppurative, 200
Norris, W. F., retinitis in Bright's disease, 192
Oertel, micrococci in nephritis, following diphtheria, 91
Passive congestion of kidney, 219
Parenchymatous nephritis, acute, 83, 85
    causes of, 85; complications of, 108;
    diagnosis of, 110; duration of, 108;
    etiology of, 85; interstitial changes
    in, 85; micrococci in, 91; morbid
    anatomy of, 88; prognosis of, 111;
    symptoms and course of, 100; syno-
    nyms of, 85; treatment of, 113; urae-
    mia in, 103; urine in, 101
Parenchymatous nephritis, chronic, 84, 124
    causes of, 124; complications of, 135;
    connective tissue, increase of, in,
    128; diagnosis of, 135; duration of,

135; etiology of, 124; morbid anatomy of, 125; prognosis of, 136; stage
of atrophy in, 128; symptoms of,
130; synonyms of, 124; treatment of,
130; urine in, 131; uremia, rarity of,
in, 133, 135
Pavy, views on diabetes, 231, 232, 234
Pyelo-nephritis, 84, 200
Rayer's views on kidney disease, 80
Reinhardt on the unity of Bright's disease, 80
Retinitis in Bright's disease, 192
    albuminuric retinitis, exceptional
    forms of, in, 195
    bibliography of, 199
    changes in the color of the fundus
    and of the retinal blood columns,
    in, 194
    curability of, 195
    forms of kidney disease in which it
    may be developed in, 195
    morbid anatomy of, 196
    statistics of, 197
    symptoms and description of, 193
    treatment of, 198
    typical cases of, 194
    uremic amaurosis of, 199
Roberts, morbid anatomy of diabetes in-
sipidus, 301
Rokitansky's views on kidney disease, 80
Rose, George H., on the arrangement of
the uriniferous tubules of the kidney, 23
Rosenstein, views on kidney disease, 81
Sclerosis, renal, 165
Schiff, experiments in producing glyco-
suria, 233-235
Stewart, classification of kidney diseases, 83
Suppurative interstitial nephritis, 84, 200
    bacteria in, 201
    calculous concretions of, in, 201
    course and duration of, 208
    diagnosis of, 213
    etiology of, 200
    infectious emboli in, 201
    morbid anatomy of, 202
    prognosis of, 214
    symptoms and course of, 205
    treatment of, 216
    urine, character of, in, 206
Surgical kidney, 200
Tube-casts, 62
    blood-casts, 62; dark granular casts,
    64; epithelial casts, 63; granular
    casts, 64; hyaline casts, 66, 67; oil-
    casts, 65, 67; pale granular casts, 64;
    waxy casts, 67
    Bartels's views on the nature of, 72;
    Beale's views, 70; Burkhardt on, 69;
    Dickinson's views on, 70, 71; John-
    son's views on, 70; Key's, Axel Ernst,
    views on, 71; Langhans's views on,
    71; Rindfleisch's views on, 73; Rob-
    in's views on, 71; Rovida's observa-
    tions on nature of, 72; Stewart's
    views on, 71; nature and clinical
    significance of, 62
Tubules of kidney, 20
Treatment of Bright's disease, etc. See In-
dex of Special Diseases.
Traube, views on kidney disease, 80, 81
Unity of Bright's disease, views on, 80, 81
Uræmia, acute, 103
    treatment of, 120
Uræmic asthma, 105
Virchow on classification of kidney dis-
ease, 81
Waxy casts, 67
    kidney, 148

Feb. 1, 1881.

CATALOGUE

OF

MEDICAL, DENTAL,

AND

SCIENTIFIC BOOKS,

PUBLISHED BY

LINDSAY & BLAKISTON.

All of which are for sale by Booksellers generally, any of which will be sent by mail, postage paid, on the receipt of the retail price, or by Express, C. O. D.

☞ Special Attention given to Filling all Orders.

LINDSAY & BLAKISTON,

MEDICAL PUBLISHERS,

25 South Sixth Street, Philadelphia.

CATALOGUE

OF THE

MEDICAL AND DENTAL PUBLICATIONS

OF

LINDSAY & BLAKISTON,

No. 25 SOUTH SIXTH STREET, PHILADELPHIA.

ACTON (WILLIAM), M.R.C.S., etc.
The Functions and Disorders of the Reproductive Organs.
Their Physiological, Social, and Moral Relations. Fourth Edition. Price 2.50

AGNEW (D. HAYES), M.D.
The Lacerations of the Female Perineum, and Vesico-
vaginal Fistula. Their History and Treatment, with numerous Illustrations. Price 1.50

AITKEN (WILLIAM), M.D.
The Science and Practice of Medicine. Third American, from
the Sixth London Edition. Containing a Colored Map, a Lithographic Plate,
and nearly 200 other Illustrations. Two volumes, royal octavo.
Bound in cloth, price 12.00; in leather, 14.00

ANSTIE (FRANCIS E.), M.D.
Stimulants and Narcotics. With Special Researches on the Action of
Alcohol, Ether, and Chloroform on the Vital Organism. Octavo. Price, 3.00

ATTHILL (LOMBE), M.D.
Clinical Lectures on Diseases Peculiar to Women. Fifth
Edition, Revised and Enlarged, with numerous Illustrations. Price 2.25

BARTH AND ROGERS.
Manual of Auscultation and Percussion. Price 1.00

BIDDLE (JOHN B.), M.D.
Materia Medica, for the use of Students. With Illustrations.
Eighth Edition, Revised and Enlarged. Price 4.00

BLACK (D. CAMPBELL), M.D., L.R.C.S., Edinburgh.
The Functional Diseases of the Renal, Urinary, and Re-
productive Organs. With a General View of Urinary Pathology.
Price 2.00

BLOXAM (C. L.).
Laboratory Teaching; or, Progressive Exercises in Practical Chemistry. Fourth Edition. Eighty-nine Engravings. Price 1.75

BYFORD (W. H.), A.M., M.D.
Practice of Medicine and Surgery. Applied to the Diseases and Accidents Incident to Women. Third Edition. Preparing.

BYFORD (W. H.), A.M., M.D.
On the Chronic Inflammation and Displacement of the Unimpregnated Uterus. A New, Enlarged Edition, with numerous Illustrations. Price 2.50

CARSON (JOSEPH), M.D.
History of the Medical Department of the University of Pennsylvania, from its Foundation in 1765. With Sketches of Deceased Professors, etc. Price 2.00

CAZEAUX (P.), M.D.
A Theoretical and Practical Treatise on Midwifery. Including the Diseases of Pregnancy and Parturition. From the Seventh French Edition, Revised and Greatly Enlarged, with numerous Lithographic and other Illustrations. Price, in cloth, 6.00; in leather, 7.00

CHARTERIS (MATTHEW), M.D.
Students' Hand-Book of the Practice of Medicine. With Microscopic and other Illustrations. Price 2.00

CHAVASSE (P. HENRY), F.R.C.S.
The Mental Culture and Training of Children, and other Subjects Relating to their Health and Happiness. Price 1.00

CLAY (CHARLES), M.D.
A Complete Hand-Book of Obstetric Surgery. With Rules of Practice in Every Emergency, and numerous Illustrations. Price 2.00

CLEAVELAND (C. H.), M.D.
A Pronouncing Medical Lexicon. Containing the Correct Pronunciation and Definition of Terms used in Medicine and the Collateral Sciences. Price, cloth, 75 cents; tucks, 1.00

COHEN (J. SOLIS), M.D.
Inhalation: Its Therapeutics and Practice. Including a Description of the Apparatus employed, etc. With Cases and Illustrations. A New Enlarged Edition. Price 2.50

COHEN (J. SOLIS), M.D.
Croup. In its Relations to Tracheotomy. Price 1.00

CORR (L. H.), M.D.
Obstetric Catechism; or, Obstetrics Reduced to Questions and Answers. With Illustrations. Price 2.00

DALBY (W. B.), F.R.C.S.
On the Diseases and Injuries of the Ear. With Illustrations.
Price 1.50

DAY (WILLIAM HENRY), M.D.
Headaches: Their Nature, Causes, and Treatment. Third Edition, Revised and Enlarged.
Price, 2.00

DILLNBERGER (DR. EMIL).
A Handy Book of the Treatment of Women and Children's Diseases.
Price 1.50

DIXON (JAMES), F.R.C.S.
The Practical Study of Diseases of the Eye. Third Edition.
Price 2.00

DUNGLISON (RICHARD J.), M.D.
The Practitioner's Reference Book. Containing Therapeutic and Practical Hints, Dietetic Rules, and General Information. Second Edition.
Price 3.50

DURKEE (SILAS), M.D.
Gonorrhœa and Syphilis. The Sixth Edition, Revised and Enlarged, with Portraits and Eight Colored Illustrations. Octavo.
Price 3.50

FENNER (C. S.), M.D., etc.
Vision: Its Optical Defects. The Adaptation of Spectacles, Defects of Accommodation, Optical Defects of the Eye, etc. Test Types and 74 Illustrations. Octavo.
Price 3.50

FENWICK (SAMUEL), M.D.
Outlines of the Practice of Medicine and Treatment of Disease. With Appropriate Formulæ and Illustrations.
Price 2.00

FOTHERGILL (J. MILNER), M.D.
The Heart: Its Diseases and Their Treatment, including the Gouty Heart. Second Edition, Entirely Rewritten and Enlarged, with Lithographic Plates and Forty other Illustrations. Octavo.
Price 3.50

FULTON (J.), M.D.
A Text-Book of Physiology. Second Edition, Revised and Enlarged, with numerous Illustrations. Octavo.
Price 4.00

GALLABIN (ALFRED LEWIS), M.D.
The Student's Guide to the Diseases of Women. With numerous Illustrations.
Price 2.00

GROSS (SAMUEL D.), M.D.
American Medical Biography of the Nineteenth Century. With Portrait of Benjamin Rush, M.D. Octavo.
Price 3.50

HABERSHON (S. O.), M.D., F.R.C.P.
On Diseases of the Stomach. The Varieties of Dyspepsia, their Diagnosis and Treatment. Third Edition. Octavo.
Price 1.75

HANDY (WASHINGTON R.), M.D.
Handy's Text-Book of Anatomy and Guide to Dissections.
For the use of Students. 312 Illustrations. Octavo. Price 3.00

HARRIS (CHAPIN A.), M.D., D.D.S.
The Principles and Practice of Dentistry. Tenth Revised Edition. In great part Rewritten, Rearranged, and with many new and important Illustrations. Edited by P. H. Austen, M.D., Professor of Dental Science and Mechanism in the Baltimore College of Dental Surgery. With nearly 400 Illustrations. Royal octavo. Price, in cloth, 6.50; in leather, 7.50

HARRIS (CHAPIN A.), M.D., D.D.S.
A Dictionary of Medical Terminology, Dental Surgery, and the Collateral Sciences. Fourth Edition, Carefully Revised and Enlarged. By Ferdinand J. S. Gorgas, M.D., D.D.S., Professor of Dental Surgery in the Baltimore College, etc. Royal octavo.
Price, in cloth, 6.50; in leather, 7.50

HAY (THOMAS), M.D.
History of a Case of Recurring Sarcomatous Tumor of the Orbit in a Child. With Illustrations. Price 50 cts.

HEADLAND (F. W.), M.D.
On the Action of Medicines in the System. Sixth American Edition, Revised and Enlarged. Price 3.00

HEATH (CHRISTOPHER), F.R.C.S.
Manual of Minor Surgery and Bandaging. For the use of House Surgeons, Dressers, and Junior Practitioners. With Formulæ and numerous Illustrations. Price 2.00

HEATH (CHRISTOPHER), F.R.C.S.
A Guide to Surgical Diagnosis. For Practitioners and Students. 12mo. Cloth. Price 1.50

HEWITT (GRAILY), M.D.
The Diagnosis, Pathology, and Treatment of Diseases of Women, Including the Diagnosis of Pregnancy. Third Edition, Revised and Enlarged, with new Illustrations. Octavo.
Price, in cloth, 4.00; in leather, 5.00

HEWSON (ADDINELL), M.D.
Earth as a Topical Application in Surgery. Being a full Exposition of its use in Cases requiring Topical Applications. With Illustrations.
Price 2.50

HIGGINS (CHARLES), F.R.C.S.
Hints on Ophthalmic Out-Patient Practice. Second Edition. 16mo. Cloth. Price .60

HILLIER (THOMAS), M.D.
A Clinical Treatise on the Diseases of Children. Octavo.
Price 2.00

HODGE (HUGH L.), M.D.
On Fœticide, or Criminal Abortion.
In paper, 30 cents: in cloth, .50

HODGE (H. LENNOX).
Note Book for Cases of Ovarian Tumers. With Diagrams, etc.
Price .50

HOFF (O.), M.D.
On Hæmaturia as a Symptom of Diseases of the Genito-Urinary Organs. Illustrated. Cloth. Price .75

HOLDEN (EDGAR), A.M., M.D.
The Sphygmograph. Its Physiological and Pathological Indications. Illustrated by Three Hundred Engravings on wood. Price 2.00

HUFELAND(C. W.), M.D.
The Art of Prolonging Life. Edited by Erasmus Wilson, M.D.
Price 1.00

JAMES (PROSSER), M.D., M.R.C.P.
Sore Throat: Its Nature, Varieties and Treatment, Including its Connection with other Diseases. Fourth Edition, Revised and Enlarged, with Colored Plates. Price 2.25

KIDD (JOSEPH), M.D., M.R.C.S.
The Laws of Therapeutics, or the Science and Art of Medicine. 12mo. Cloth. Price 1.25

KOLLMEYER (A. H.), A.M., M.D.
Chemia Coartata; or, The Key to Modern Chemistry. With Numerous Tables, Tests, etc., etc. Price 2.25

LAWSON (GEORGE) F.R.C.S.
Diseases and Injuries of the Eye, their Medical and Surgical Treatment. Containing a Formulary, Test Types, and Numerous Illustrations. Fourth Edition. Preparing.

LEBER AND ROTTENSTEIN (Drs.)
Dental Caries and Its Causes. An Investigation into the Influence of Fungi in the destruction of the Teeth. With Illustrations. Price 1.25

LEWIN (Dr. GEORGE).
The Treatment of Syphilis by Subcutaneous Sublimate Injections. With a Lithographic Plate, etc. Price 1.50

LIZARS (JOHN), M.D.
The Use and Abuse of Tobacco. Price .50

LONGLEY (ELIAS).
Student's Pocket Medical Lexicon. Giving the Correct Pronunciation and Definition of all Words and Terms in general use in Medicine and the Collateral Sciences. Cloth, 1.00; leather, with tucks and pocket, 1.25

MAXSON (EDWIN R.), M.D.
The Practice of Medicine. Price 3.00

MAYS (THOMAS J.), M.D.
On the Therapeutic Forces; or, The Action of Medicine in the Light of the Doctrine of Concervation of Force. Price 1.25

MEADOWS (ARFRED), M.D.
　A Text-Book of Midwifery. Including the Signs and Symptoms of Pregnancy, Obstetric Operations, Diseases of the Puerperal State, etc. Second Edition, Revised and Enlarged. 145 Illustrations. 　Price 3.00

MEARS (J. EWING), M.D.
　Practical Surgery. Including Surgical Dressings, Bandaging, Amputation, etc., etc., with 227 Illustrations. 　Price 2.00

MEIGS AND PEPPER.
　A Practical Treatise on the Diseases of Children. By J. Forsyth Meigs, M.D., Fellow of the College of Physicians of Philadelphia, etc., etc., and William Pepper, M.D., Physician to the Philadelphia Hospital, etc. Sixth Edition, thoroughly Revised and Enlarged. A Royal Octavo Volume of over 1000 pages. 　Price, bound in cloth, 6.00; leather, 7.00

MACKENZIE (MORELL), M.D.
　Diphtheria: Its Nature and Treatment. 12mo. 　Price .75

MENDENHALL (GEORGE), M.D.
　Medical Student's Vade-mecum. A Compendium of Anatomy, Physiology, Chemistry, the Practice of Medicine, Surgery, Obstetrics, etc., etc. Eleventh Edition. 224 Illustrations. In cloth. 　Price 2.00

MILLER (JAMES), F.R.C.S.
　Alcohol, Its Place and Power. Cloth. 　Price .50

MILLER & LIZARS.
　Alcohol: Its Place and Power; and The Use and Abuse of Tobacco. The Two Essays in One Volume. Cloth. 　Price 1.00

NORRIS (GEORGE W.), M.D.
　Contributions to Practical Surgery, including numerous Clinical Histories, etc. 　Price 4.00

OTT (ISAAC), M.D.
　The Actions of Medicines. With Twenty-two Illustrations. 　Price 2.00

PENNSYLVANIA HOSPITAL REPORTS.
　Edited by a Committee of the Hospital Staff. J. M. Da Costa, M.D., and William Hunt, M.D. Vols. 1 and 2, containing Original Articles, by former and present Members of the Staff, with Lithographic and other Illustrations. 　Price per volume 2.00

PEREIRA (JONATHAN), M.D., F.R.S., etc.
　Physician's Prescription Book. Containing Lists of Terms, Phrases, Contractions and Abbreviations used in Prescriptions, Explanatory Notes. Grammatical Construction of Prescriptions, Rules for the Pronunciation of Pharmaceutical Terms, etc., etc. Sixteenth Edition.
　　Price, in cloth, 1.00; in leather, with tucks and pocket, 1.25

PHYSICIAN'S VISITING LIST, PUBLISHED ANNUALLY.

SIZES AND PRICES.

For 25 Patients weekly.	Tucks, pockets, and pencil,						1.00
50 " "	" " "						1.25
75 " "	" " "						1.50
100 " "	" " "						2.00
50 " " 2 vols.	{ Jan. to June } { July to Dec. } "						2.50
100 " " 2 vols.	{ Jan. to June } { July to Dec. } "						3.00

INTERLEAVED EDITION.

For 25 Patients weekly, interleaved, tucks, pockets, etc.,						1.25
50 " " " " " "						1.50
50 " " 2 vols. { Jan. to June } { July to Dec. } "						3.00

This Visiting List, now in its twenty-ninth year, contains the Metric or French Decimal System of Weights and Measures, a Posological Table with the Doses in both the Apothecaries' and Decimal Metric System of Weights and Measures, a new Table of Poisons, etc.

PIGGOTT (A. SNOWDEN), M.D.
Copper Mining and Copper Ore. With a full Description of the Principal Copper Mines of the United States, the Art of Mining, etc. Price 1.00

POWER, HOLMES, ANSTIE, AND BARNES.
Reports on the Progress of Medicine, Surgery, Physiology, Midwifery, Diseases of Women and Children. Price 2.00

PRINCE (DAVID), M.D.
Plastic and Orthopedic Surgery. Containing a Report on the Condition of, and Advance made in, Plastic and Orthopedic Surgery, etc., etc., and numerous Illustrations. Price 4.50

RADCLIFFE (CHARLES BLAND), M.D.
On Epilepsy, Pain, Paralysis, and other Disorders of the Nervous System. With Illustrations. Price 1.50

REESE (JOHN J.), M.D.
Analysis of Physiology. A Condensed View of the most important Facts and Doctrines for Students. Second Edition, Enlarged. Price 1.50

REESE (JOHN J.), M.D.
The American Medical Formulary. Price 1.50

REESE (JOHN J.), M.D.
A Syllabus of Medical Chemistry. Price 1.00

REYNOLDS (J. RUSSELL), M.D., F.R.S.
Lectures on the Clinical Uses of Electricity. Second Edition. Price 1.00

RICHARDSON (JOSEPH), D.D.S.
A Practical Treatise on Mechanical Dentistry. Third Edition, much Enlarged. With 185 Illustrations. Octavo. Cloth, 4.00; leather, 4.75

RIGBY AND MEADOWS.
Obstetric Memoranda. Fourth Edition, Revised and Enlarged, by ALFRED MEADOWS, M.D.   Price .50

RINDFLEISCH (DR. EDWARD).
Text-Book of Pathological History. Translated from the German, by WM. C. KOLMAN, M.D., assisted by F. T. MILES, M.D. 208 Microscopical Illustrations. Octavo.   Price, bound in cloth, 5.00; in leather, 6.00

ROBERTS (FREDERICK T.), M.D., B.Sc.
The Theory and Practice of Medicine. Third American, from Fourth London Edition. Revised and Enlarged. With Illustrations.
Price, cloth, 5.00; leather, 6.00

ROBERTS (D. LLOYD), M.D.
The Student's Guide to the Practice of Midwifery. With 95 Illustrations.   Price 2.00

RYAN (MICHAEL), M.D.
Philosophy of Marriage, in its Social, Moral, and Physical Relations; the Diseases of the Genito-Urinary Organs, etc.   Price 1.00

SANDERSON, KLEIN, FOSTER, AND BRUNTON.
A Hand-Book for the Physiological Laboratory. Being Practical Exercises for Students in Physiology and Histology. With over 350 Illustrations, appropriate letter-press explanations and references to the text. 2 vols., cloth, $7.00. Price, in one vol., cloth, 6.00; in leather, 7.00.

SANSOM (ARTHUR ERNEST), M.B.
Chloroform. Its Action and Administration.   Price 1.50

SEWILL (H. E.), M.R.C.S., Eng., L.D.S.
The Student's Guide to Dental Anatomy and Surgery. With 77 Illustrations.   Price 1.50

SMITH (HEYWOOD), M.D.
Practical Gynæcology. A Hand-Book for Students and Practitioners. With Illustrations.   Price 1.50

STILLÉ (ALFRED), M.D.
Epidemic Meningitis; or, Cerebro-spinal Meningitis.   Price 2.00

STOKES (WILLIAM).
The Diseases of the Heart and the Aorta. Octavo.   Price 3.00

SWAIN (WILLIAM PAUL), F.R.C.S.
Surgical Emergencies: Containing Concise Descriptions of Various Accidents and Emergencies, with Directions for their Immediate Treatment. With Illustrations.   Price 2.00

SWERINGEN (HIRAM V.), M.D.
A Pharmaceutical Lexicon; or, Dictionary of Pharmaceutical Science. Containing explanations of the various subjects and terms of Pharmacy, with appropriate selections from the collateral sciences. Formulæ for officinal, empirical, and dietetic preparations, etc., etc. Price, in cloth, 3.00; in leather, 4.00

TAFT (JONATHAN), D.D.S.
A Practical Treatise on Operative Dentistry. Third Edition, thoroughly Revised, with Additions. Over 100 Illustrations. Octavo.
Price, in cloth, 4.25; in leather, 5.00

TANNER (THOMAS HAWKES), M.D., F.R.C.P., etc.
The Practice of Medicine. Sixth American Edition. Revised and Enlarged. With extensive Formulæ for Medicines, Baths, etc., etc. Royal Octavo; over 1100 pages. Price, in cloth, 6.00; leather, 7.00

TANNER (THOMAS HAWKES), M.D., F.R.C.P., etc.
Index of Diseases and their Treatment. Second Edition. Carefully Revised. With many Additions and Improvements. By W. H. BROADBENT, M.D., F.R.C.P. Octavo. Cloth. Price 3.00

TANNER (THOMAS HAWKES), M.D., F.R.C.P., etc.
A Practical Treatise on the Diseases of Infancy and Childhood. Third Edition, Revised and Enlarged. By ALFRED MEADOWS, M.D. Price 3.00

TANNER (THOMAS HAWKES), M.D., F.R.C.P., etc.
A Memoranda of Poisons. Fourth Edition, much Enlarged. Price .75

TIBBETS (HERBERT), M.D.
A Hand-Book of Medical Electricity. Sixty-four Illustrations.
Price 1.50

TOLAND (H. H.), M.D.
Lectures on Practical Surgery. Second Edition. With Additions and numerous Illustrations. Price, in cloth, 4.50; in leather, 5.00

TRANSACTIONS OF THE COLLEGE OF PHYSICIANS OF PHILADELPHIA.
New Series, Vols. I, II, III, and IV. Price, per volume, 2.50

TROUSSEAU (A).
Lectures on Clinical Medicine. Delivered at the Hôtel Dieu, Paris. Translated from the Third Revised and Enlarged Edition by P. VICTOR BAZIRE, M.D., and JOHN ROSE CORMACK, M.D. With Index, Table of Contents, etc. Complete in Two Volumes, royal octavo.
Price, bound in cloth, 8.00; in leather, 10.00

TYSON (JAMES), M.D.
A Treatise on Bright's Disease and Diabetes. With Especial Reference to Pathology and Therapeutics. In Press.

TYSON (JAMES), M.D.
The Cell Doctrine. Its History and Present State, with a Copious Bibliography of the Subject. With Colored Plate, and numerous other Illustrations. Second Edition. Price 2.00

TYSON (JAMES), M.D.
A Practical Guide to the Examination of Urine. For the Use of Physicians and Students. With a Colored Plate and numerous other Illustrations. Third Edition. Price 1.50

TURNBULL (LAURENCE), M.D.

The Advantages and Accidents of Artificial Anæsthesia.
A Manual of Anæsthetic Agents, Modes of Administration, etc. Second
Edition, Enlarged. 25 Illustrations. Cloth.   Price 1.50

WALKER (ALEXANDER), M.D.

Intermarriage; or, the Mode in which, and the Causes why, Beauty, Health
and Intellect result from certain Unions; and Deformity, Disease and Insanity
from others. With Illustrations. 12mo.   Price 1.00

WARING (EDWARD JOHN), F.R.C.S., etc.

Practical Therapeutics. Considered chiefly with reference to Articles
of the Materia Medica. Third American, from the last London Edition.
                              Price, in cloth, 4.00; leather, 5.00

WEDL (CARL), M.D.

Dental Pathology. With Special Reference to the Anatomy and Physiology of the teeth, and notes by Thos. B. Hitchcock, M.D., Prof. of Dental
Pathology, Harvard University. 105 Illustrations.
                              Price, cloth, 3.50; leather, 4.50

WHEELER (C. GILBERT), M.D.

Medical Chemistry: Including the Outlines of Organic and Physiological
Chemistry, Second Edition. Cloth.   Price 3.00

WILSON (JOSEPH), M.D.

Naval Hygiene; or, Human Health and means of Preventing Disease,
With Illustrative Incidents derived from Naval Experience. Illustrations,
etc. Second Edition.   Price 3.00

WOODMAN AND TIDY.

A Text-Book of Forensic Medicine and Toxicology. By
W. Bathurst Woodman, M.D., Assistant Physician and Lecturer on Physiology, London Hospital; and C. Meymott Tidy, M.A., M.B., Lecturer on
Chemistry, and Professor of Medical Jurisprudence, London Hospital. Numerous Illustrations.   Price, cloth, 7.50; leather, 8.50

WYTHE (JOSEPH H.), A.M., M.D.

The Physician's Pocket Dose and Symptom Book. Contain-
ing the Doses and Uses of all the Principal Articles of the Materia Medica, and
Original Preparations. Eleventh Revised Edition.
                    Price, in cloth, 1.00; in leather, tucks, with pockets, 1.25

WYTHE (JOSEPH H.), A.M., M.D.

The Microscopist: A Manual of Microscopy and Compendium of the
Microscopic Sciences, Micro-Mineralogy, Micro-Chemistry, Biology, Histology,
and Practical Medicine. Fourth Edition, 252 Illustrations.
                              Cloth, 5.00; leather, 6.00

OVERMAN (FREDERICK), M.E.

Practical Minerology, Assaying, Mining. Tenth Editon.
                              Cloth, 1.00 .

MATTHIAS (BENJAMIN), A.M.

A Manual for conducting Business in Town and Ward
Meetings Sixteenth Edition.   Cloth, price, 50

www.ingramcontent.com/pod-product-compliance
Lightning Source LLC
Chambersburg PA
CBHW060531030726
47498CB00004B/1149